AGRICULTURAL MARKETING IN GUJARAT

DR. ANITA ARYA obtained her B.A. degree from St. Xaviers College Ahmedabad in 1976. Thereafter, she obtained M.A. from Gujarat University in 1978 and specialised in Agricultural Economics. At M.A. she ranked second in the University in the subject of Economics. After her M.A. she joined Indian Institute of Management, Ahmedabad and worked on several projects, and simultaneously she taught Economics in St. Xaviers College, Ahmedabad. In 1990, she obtained her Ph.D. from Gujarat University. At present she is working at Sardar Patel Institute of Economics and Social Research, Ahmedabad. She has participated in number of seminars and conferences.

AGRICULTURAL MARKETING IN GUJARAT

ANITA ARYA
Sardar Patel Institute of Economic & Social Research
Ahmedabad

CONCEPT PUBLISHING COMPANY, NEW DELHI 110059

ISBN 81-7022-449-7

First Published 1993

Published and Printed by
Ashok Kumar Mittal
Concept Publishing Company
A/15-16, Commercial Block, Mohan Garden
NEW DELHI-110059

Lasertypeset by
Microtech Advance Printing Systems (Pvt.) Ltd.
H-13, Bali Nagar
NEW DELHI-110015

PREFACE

I am deeply grateful to my guide Dr. P.K. Dave, Department of Economics, Gujarat University for guiding me in my present research work. His constant encouragement made me to work tirelessly with an exclusive goal of the completion of my work. Further, he made himself readily available for consultation and discussion and for that reason only, I could complete my work within reasonable time limit.

I owe my sincere thanks to Dr. R.J. Mody, Director, Sardar Patel Institute of Economic and Social Research for giving me all necessary facilities in the institute for the furtherance of my research. I am also thankful to Prof. N.C. Shah, Associate Professor, Sardar Patel Institute of Economic and Social Research for guiding me in the statistical analysis. Further, I thank all the senior faculty members and colleagues of Sardar Patel Institute of Economic and Social Research for showing concern in the completion of my work.

I am thankful to Prof. V.P. Bhardwaj, Professor and Head, Department of Economics, Gujarat University for making valuable suggestions to introduce qualitative improvements in the study. I also thank my teachers especially Dr. C.H. Thaker and Prof. P.C. Vora of the Department of Economics, Gujarat University who have shown positive interest in the completion of my work. I owe my sincere thanks to Shri D.S. Parmar of Indian Institute of Management, Ahmedabad for taking keen interest in the last stages of my work. I cannot forget my senior friend Smt. Manjulaben Naik who has given me inspiration throughout the period of my research work. I shall be failing in my duty if I do not express my thanks to my parents who have given me constant encouragement and moral support in my research work.

I am happy to put my doctoral dissertation work in the form of a book. While publishing it, an attempt has been made to revise and abridge the original work. Some of the tables of the original work have been rearranged and presented in a more meaningful way especially in Chapter V. Whereas, some of the tables from Chapter IV have been removed when it has no bearing on the quality of the discussion. Similarly, some portions like parity prices have been shifted to the Appendix. The graphs pertaining to fluctuations in prices and trends in income of the Agricultural Produce Market Committees, have also been shifted to the Appendix for the purpose of convenience.

While publishing my research work, I am again thankful to my guide, Dr. P.K. Dave, who spared his time in discussing again the various aspects of my study.

I am also thankful to Dr. U.K. Srivastava, Professor, Centre for Management in Agriculture, Indian Institute of Management, Ahmedabad, who gave me constant inspiration for the publication of my doctoral dissertation.

ANITA ARYA

June 1991

CONTENTS

LIST OF TABLES

I

MEANING, DEFINITION AND HISTORY OF AGRICULTURAL MARKETING

Introduction

Marketing is a technique of buying and selling. There are two sets of people involved in marketing - those who produce economic goods and have surplus and those who need these surpluses. Marketing postulates favourable and conducive conditions in which these two sets of people by "Exchange" fulfil their mutual demand. Marketing includes buying, selling, transport, storage, finance, risk-bearing, standardisation etc.

The study of Agricultural Marketing involves the study of all the agencies and steps through which a commodity passes till it reaches the final consumer. In the process of marketing, assembling, grading, standardisation, storage, transport, wholesaling and retailing play their prescribed roles and there are various agencies performing them. According to the National Commission on Agriculture (XII Report), agricultural marketing is a process which starts with a decision to produce agricultural commodity and it involves all the aspects of market system viz. functional and institutional based on economic and technical considerations and includes pre and post harvest operations, assembling, grading, storage, transportation and distribution. All activities involved from the time agricultural product leaves the farmer and reaches the consumer are considered as agricultural marketing by Kohls[1].

Agricultural marketing includes all services tangible and intangible rendered between the farmers and the consumers. Pricing, financing, transporting, processing, grading etc., are included in the activities of agricultural marketing.[2]

Agriculture is the backbone of Indian economy and in order that agriculture plays a vital role in our economic development, its production should increase. The farmer has to increase marketable surplus and here lies the importance of agricultural marketing.

But due to his illiteracy, ignorance financial weakness, the farmers possess a very weak bargaining power and on the other hand the trader takes the advantage of the farmer's weakness because he is generally well informed, well organised, financially sound and very tactful in his dealings. The result of all this is that the farmer remains at the mercy of the trader through generations. The market also is deprived of healthy and fair competition. The farmer is discontended with the environment in which he has to operate in the market. Some of the unfair deals and undesirable practices viz. short weights, considerable delay in payment, high market charges, taking away substantial quantity under the name of sample, underhand dealings, the incorrect and false weights are fairly known in agricultural marketing in India. Also one can never forget the customary gesture of secret or cover bids.

The main defects of agricultural marketing in India are the following:

1. Defective preparation for marketing
2. Lack of organisation among cultivators
3. Forced sale in villages
4. Superfluous middlemen
5. Multiplicity of market charges
6. Multiplicity of weights and measures
7. Adulteration
8. Malpractices in the market
9. Absence of market intelligence
10. Absence of grading and standardisation
11. Lack of transport facilities
12. Lack of storage facilities

We all are well versed with the above mentioned defects of the agricultural marketing system. There is hardly any stage of marketing which is not suffering from one defect or the other.

Thus, the trader has the upper hand in the agricultural marketing system and the farmer is always at his mercy.

Meaning and Definition

Government interference in the field of agricultural marketing is a must. In fact, if one party exploits the other in the game of agricultural marketing, government intervention becomes inevitable.

The government interference in the process of marketing brings about a conducive environment of trust and confidence between the buyers and the sellers. The governmental intervention postulates fair trade policies and assures both the parties fair and reasonable returns.

As explained by Rhodes, Tomek and Robinson and Dahl and Hammond[3], organised markets are public markets, which have a set of formal rules or customs to regulate the time and place of trading, the obligations of each party, the mode of transportation, and similar matters. Their operation is quite public. Prices are publicly arrived at.

A market is said to be regulated when the state government establishes a market under some enactment and lays down certain rules and regulations to carry out the business of marketing.

According to Gupta[4], a market is said to be regulated when the state government establishes a market under some enactments and frame rules and regulations to conduct business therein.

As defined by Johnson[5], market regulation comprises a set of acts deliberately designed by the state as a protective shield to farmers against traders' capacity.

Regulated market is an agency which brings a radical change in the interest of cultivator first and those engaged in trade and commerce next.

The chief objective of Regulated Market is to regulate buying and selling, encourage fair and healthy competition by creating conditions for the same, and hence both the buyer and the seller gets fair return in the deal. The objective apparently is to create structural conditions necessary for effective and allocatively efficient competition. It is assumed that market regulations can curb monopolistic tendencies in the system, and 'efficiency' and 'equity' can be restored.[6]

Thomson[7] describes the objectives of the agricultural marketing system into three types which the different interests such as farmers, middlemen and consumers desiré for better marketing. The consumer desires a marketing system which will help to buy his requirement at the lowest possible cost. The middlemen wants the largest possible returns for his products and the farmer wants a marketing system which ensures him the minimum costs and the highest possible prices for the products. Further, according to Wong[8], the objective of agricultural marketing is to ensure that sales takes place as quickly as possible and that no time is wasted.

Market regulation is a socio-economic reform, hence cultivators are to be educated about improvements in the marketing of their produce. They should also be conscious of their rights in the management and control of the market and should not submit meekly to questionable practices of the commission agents.

In the regulated market the sale of agricultural commodities is done by open auction system or by the close tender method. The sale is done under the supervision of an official of the market committee and the trader has to give his signature as soon as the auction is over. The transaction is done during fixed hours. The farmer has the option to withdraw his produce even at the highest bid, if he so desires.

Weighment is done by the licenced weighmen with standard weights and platform scale. In some markets, a weigh bridge has been installed to avoid short weight and malpractices.

Grading of produce is a must but according to the National Commission on Agriculture, only 13 per cent of the regulated markets have grading facilities.

The loud speakers and notice boards provide information about prices and other related things.

The marketing charges are returns to retailers, wholesalers, processors and assemblers for their services in the marketing of agricultural products. The market charges include commission charges, brokerage, hamali and weighing charges. The seller has to pay for transportation and hamali only. The rest is borne by the buyer.

The farmer receives his money payment of his produce sold to the trader immediately without any delay.

All the market functionaries from the hamals to the traders have to obtain a licence, after paying the prescribed fees, to carry on their business. The licenced traders keep up to date records and accounts in accordance with the bye-laws. The middlemen are subject to punishment, if they violate the rules and regulations.

The functioning of the regulated markets is supervised by the secretary, auction clerk and other staff members of the market committee.

The market committee consists of farmers, traders, cooperative marketing societies, cooperative or commercial banks, autonomous bodies (Panchayat Samiti and Municipal Board of the area) and government officials.

The market committee is nominated for three years and the members of the committee elect their chairman and deputy chairman.

The chief source of funds of the market committee are:

(i) Market fee on the produce brought for sale in the market yard/market area. The rate of market fee varies from state to state ranging from Rs. 0.20 to Rs.0.30 for the commodities worth Rs.100.

(ii) Licence/renewal fee of middlemen

(iii) Subsidy from the government

The sub-committee of the market committee is expected to solve the disputes among producer-seller and traders.

The regulated markets have eliminated the superfluous middlemen. The sale slips of the produce sold are given to the producer-seller, giving the details of the quantity sold, the price offered and the deductions. A copy of the sale slip is given for checking to the market committee.

The market committee provides various amenities like spacious market yard, rest house, cattle-shed and water troughs, lighting, watchman, drinking water, parking space for carts and other infrastructure facilities like bank, post office, canteen etc. in the yard.

Thus, the farmer enjoys quite a number of benefits in doing business in a Regulated Market. He has his own representation in the management of the market committee. The other advantages are open bid system of sale, standardised weights and market charges, immediate and full payment. Last but not the least is the protection from the middlemen.

The Regulated Markets have developed a 'Corporate Sense' among the producers. The farmers have become very much informed, alert and have acquired a lot of trust and confidence in the regulated markets. The farmers are the part of the management and their association with the administration have made them aware of the advantages of different marketing reforms and above all they have now, started understanding the exploitative traders.

Agricultural Marketing in Gujarat

The Gujarat State has an area of 195,984 sq.kms. Its population is 3.39 crores and 69 per cent of which is the rural population. In the state of Gujarat, 41 lakh farmers and farm workers are engaged in the primary sector which forms 38 per cent of the total work force in the state.

There are 19 districts which are divided into 184 talukas.

Bajra, wheat, rice, and jowar are the most important food crops; while groundnut, cotton and tobacco are important cash crops. Gujarat ranks fourth after Punjab, Haryana and Maharashtra in terms of per capita state domestic product at current prices in 1977-78 and 1978-79[9]. In Gujarat, during 1960-61 to 1980-81, the compound growth rate of State domestic product (at 1970-71 constant price) was 3.5 per cent per annum. The primary sector contributed 27 per cent to the State domestic product in the bad years and 42 to 44 per cent in really good years. However, the compound growth rate of the primary sector in the State was 2.4 per cent per annum which was lowest as compared to the secondary (4.1 per cent per annum) and tertiary (4.3 per cent per annum) sectors.

In Gujarat between 1960-61 and 1977-78, the gross cropped area increased from 97.68 lakh hectares to 104 lakh hectares, the gross irrigated area increased from 7.34 lakh hectares to 19.36 lakh hectares. During 1960-61 to 1979-80 yield per hectare of wheat, jowar and bajra recorded 4 to 5 per cent compound growth rate per annum and the compound growth rate of groundnut and cotton was 2 per cent per annum and that of tobacco was 4 per cent. The most important feature of Gujarat agriculture is the rapid growth in the production of both food and nonfood crops The farm production has increased over the years. For example, in 1960, the foodgrain production was only 26.25 lakh M.T. which has gone up to 46.19 lakh M.T. in 1984-85 Similarly, the production of commercial crops like groundnut, tobacco, condiments has also gone up substantially.

Gujarat has the distinction of having the largest number of farmers' marketing and processing cooperatives which function effectively. At the state level, there are five commodity marketing federations and a general marketing federation At regional and taluka levels, there are large number of commodity marketing cooperatives. Producers' cooperatives directly serve the interest of farmers. Integrated cooperatives are formed wherever processing of commodities is involved. Cooperative institutions are more active in commercial crops than they are in foodcrops.

When Gujarat State came into existence, Bombay Agricultural Produce Act of 1939 and Saurashtra Agricultural Produce Market Act 1955 were in force in respective areas. No regulation was in force in Kutch. State Government experienced difficulties in implementing two different Acts and there was no Act for Kutch Area. Moreover, some of the provisions of the Act were challenged in the court. Therefore, it was necessary on the part of the state to remove these difficulties and to cover the whole state of Gujarat under one Act. As a result, Gujarat government appointed a committee, under the chairmanship of Jaswantlal Shah to make a comprehensive study of the problem. On the basis of the recommendations of the committee, present Agricultural Produce Marketing Act came into force from 1st June, 1964 Though the Act was progressive, many loopholes were found in it Therefore, a new committee was appointed under the chairmanship of Maldevjibhai Odedara with a view to make it more effective in achieving socialist objectives The committee submitted its report on 22nd December 1975 to make necessary changes on the basis of the recommendations of Odedara committee After that, a new committee was appointed on 4th January 1983 under the chairmanship of Shri Shankarlal Guru.

The market area consists of a taluka and the market proper consists of villages within a radius of 10 to 15 km of the market yard Licence fees to the traders, commission agents, brokers, carting agents, weighmen, hamals are fixed by the market committees under the bye-laws The licence fees are subject to minimum and maximum prescribed by the Gujarat Agricultural Produce Marketing Rules, 1965 The state government fixes the minimum market fee and this can be revised by the market committee Initially the farmers used to pay the market fee but since 1982-83, the government of Gujarat has made changes in the Act whereby the traders bear the burden of the market fee.

The market committee (a corporate body) has producers, traders, cooperatives, government nominees, local bodies etc. in its management Moreover, the producers/sellers have more than 50 per cent representation in the market committee and there is a special and specific provision for this in the Act.

The composition of the Market Committee in Gujarat State for each market is as under[10].

However 17 members constitute the market committee Thus the market committee consists of growers, traders, local institutions, cooperative bodies. The committee is headed by the Chairman and a Vice Chairman and the term of the committee is of three years. The secretary of the committee executes the market and all the employees are directly responsible to him The secretary is employed by the market committee and he is confirmed by the department of cooperatives He plays a crucial role in the functioning of the market and his tenure depends on his performance and conduct The market committee is responsible to pay the salaries of the Secretary and the other staff members.

Constituency	Number of Members
(1) Growers	8
(2) Representative of traders	4
(3) Local bodies	1
(4) Nominated from cooperative bodies	2
(5) Appointed or nominated by government	2
(6) Total	17

The market committee is supposed to maintain and manage market, prevent adulteration, promote grading and standardisation and implement the provisions of the Act, Rules, bye-laws and the conditions of the licences. The market committee is the local authority and has to sanction its budget, appoint its staff except Secretary, issue licences and levy and collect market fees.

The Functions of the Market Committee

1. Maintaining and managing the market (Sec.26).
2. Enforcing in the market area, the provisions of the Act, the Rules and the Bye-laws (Sec.26).
3. Enforcing the conditions of licences granted to different market functionaries in connection with the purchase and sale of agricultural produce regulated by it (Sec.26).
4. Granting, renewing, suspending and cancelling a general or a special licence for the purchase and sale of regulated agricultural produce.
5. Maintaining the register containing the names of all licenced traders and general commission agents (Sec.27).
6. Providing all facilities in the market as per the directions of the Director (Sec.26).
7. Preparing budget estimate of its income and expenditure for the next succeeding year and sanctioning it (with or without modification) within the prescribed period (Rule 42).

8. Levying fees on the agricultural produce bought or sold in the market area and collecting them through agents or otherwise (Sec.28).
9. Managing the income, incurring the expenditure and investing the surplus funds (Sec. 32 and 33).
10. Keeping the account and the other records in such a manner as directed by the Director or any other officer authorised by him on his behalf (Rule 47).
11. Maintaining a register showing the fees collected by it (Rule 50).
12. Paying contribution every year, at prescribed rates to the State Agricultural Produce Fund (Sec. 34).
13. Prescribing the market charges, with the previous sanction of the Director, for the services of different market functionaries rendered in connection with the purchase and sale of agricultural produce regulated (Bye-law 31).
14. Taking all possible steps to prevent adulteration of agricultural produce in the market area, including the prevention of the mixing of the last picking of cotton or inferior variety of cotton with the superior one, bold linseed with small linseed, ghee with vegetable oil and of earth, dirt, stones or any other extraneous matter with any other agricultural produce (Sec.16 and Rule 73).
15. Promoting, grading and standardisation of agricultural produce by taking such steps as (a) maintaining an up to date set of samples of standard grades of agricultural produce sold at the market for use of the sellers and the buyers (b) keeping and exhibiting samples of different grades of agricultural produce with indications of parity prices based on rates ruling in terminal and key markets for the information of the sellers and the buyers, (c) undertaking the grading of agricultural produce, and (d) carrying out or supervising the ginning of pure varieties of cotton brought into the market (Sec.26 Rule 74).
16. Collecting and maintaining daily list of prices of different types and grades of agricultural produce regulated by it and supplying them to the Government when required (Sec.26).
17. Placing information on matters like the prices of the commercial crops ruling at the principal marketing centres of the tract and at the ports at the disposal of the users of the market by publishing it (Rule 55).

18. Communicating the full names and addresses of the traders holding general licences in the market area to the authorised officer within the prescribed time when general election or by-election is to be held (Rule 7).
19. Prescribing the units of quotations in respect of the regulated agricultural produce to be bought and sold (Bye-law 33).
20. Appointing the secretary of the committee (with the approval of the Director) and other officers and servants for managing its affairs (Sec.22).
21. Appointing a Disputes Committee for the purpose of settlement of disputes that may arise between the buyers and the sellers (Rule 40).
22. Appointing one or more sub-committees for carrying out any work or to resort to it on any matter and delegating such of its powers and duties to it or them or to any of its members as may be deemed necessary (Sec.25), and
23. Issuing all directions necessary for the guidance of the persons using the market by publishing them on the notice board (Bye-law 54).

In Gujarat, there are 350 mandi bazars where the farmers bring their agricultural produce for sale. Table 1.1 gives districtwise number of market committees, market yards and sub-yards in Gujarat state during 1980-81.

In Gujarat there are 149 market committees and total number of market yards are 321 out of which 131 are main market yards and 190 sub-market yards.

The district of Mehsana has the maximum number of main market yards and sub-market yards, i.e. 12 and 21 respectively, followed by Kheda district (11 and 11 respectively) Mehsana and Kheda have the maximum number of market committees also, i.e. 12 and 11 respectively.

Junagadh district has the minimum number (6) of Agricultural Produce Market Committees and it has only 3 main yards. Kutch district also has 6 market committees but it has 5 main yards. In Dang district, there is not a single market yard. In Gandhinagar district of Gujarat state, there is only one market committee and one main market yard. This is because it is a single taluka district and has a small area of 649 sq.kms.

TABLE 1.1: Districtwise Number of Market Committees, Market Yards and Sub-Market Yards in Gujarat

Sr. No.	Name of the District	No. of APMC	No.of Main Yard	No. of Sub-Yard
1	Ahmedabad	7	7	13
2	Gandhinagar	1	1	-
3	Sabarkantha	9	9	17
4	Mehsana	12	12	21
5	Banaskantha	9	9	9
6	Kheda	11	11	22
7	Panchmahal	10	10	23
8	Vadodara	8	8	22
9	Bharuch	8	8	17
10	Surat	10	10	23
11	Dang	-	-	-
12	Valsad	8	6	15
13	Rajkot	8	8	5
14	Jamnagar	7	4	-
15	Kutch	6	5	-
16	Junagadh	6	3	-
17	Amreli	9	7	1
18	Surendranagar	9	5	-
19	Bhavnagar	10	7	2
20	Gujarat State Tobacco Committee	1	1	-
	Total	149	131	190

Source: Gujarat State Agricultural Marketing Board, Gandhinagar.

The market committees are rapidly growing in number and are becoming financially strong The table 1.2 shows the income and expenditure of the market committees and the table 1.3 shows the number of licenced traders.

TABLE 1.2: Income and Expenditure of Market Committees

Year	No. of Market Committee	Income (Rs. lakhs)	Expenditure (Rs lakhs)	Surplus Amount	Permanent Fund
1980-81	144	316.58	204.94	111.93	943.63
1981-82	145	394.33	252.01	148.61	981.63
1982-83	147	518.49	294.10	215.22	1194.88
1983-84	148	776.52	356.71	376.69	1323.80
1984-85	148	773.64	439.40	334.24	1780.46
1985-86	149	858.78	527.45	331.33	2257.90

TABLE 1.3: Number of Licence Holder

Year	No. of Licence traders	No. of Co-ops. having licences of market	Total No. of licence holders Committee
1980-81	109133	1774	110907
1981-82	135695	2084	117779
1982-83	113298	2109	115497
1983-84	118704	2056	120760
1984-85	109552	2099	111651
1985-86	110931	2019	113110

The market arrivals of important agricultural crops of four years are given in Table 1.4.

The grading at the producer's level is done at Dahod, Bodeli, Unjha, Patan, Mahuva, Bilimora, Sami, Talod and Gondal market. The agricultural commodities that are graded in these markets are viz. groundnut, wheat, bajra, gram, chikku, cotton etc. Cumin seed is graded at Unjha and Patan; whereas ghee, vegetable oils and ground spices are graded under Agmark at Veraval, Khambhalia, Porbandar and Bareja market centres.

TABLE 1.4: Market Arrivals of Important Crops of Regulated Markets of Gujarat During 1982-83 to 1985-86

(In lakhs quintals)

Sr.No.	Name of Commodities			
	1982-83	1983-84	1984-85	1985-86
1. Cotton	52.45	40.74	48.82	52.64
2. Groundnut	16.21	23.88	41.21	10.75
3. Paddy	36.62	51.25	46.60	64.66
4. Wheat	49.68	65.06	55.85	52.03
5. Bajra	27.48	21.72	17.29	16.85
6. Jowar	21.72	9.76	11.40	33.74
7. Oilseeds (other than groundnut)	54.34	54.46	82.09	57.00
8. Pulses	33.55	28.27	37.79	26.48
9. Isabgul	3.00	6.68	6.69	5.44
10. Chillies (Dry)	0.85	0.81	0.76	0.31
11. Fruits & Veg. & Coconut	99.27	101.96	78.26	119.87
12. Tobacco	2.11	1.84	1.54	0.46
13. Grass	1.89	1.64	2.19	1.60
14. Rest	35.58	48.49	35.12	20.69
Total	431.75	468.55	465.61	462.52

For the benefit of the market functionaries, a fortnightly bulletin of market arrivals and prices is published in the local language viz. Gujarati. The All India Radio Stations at Vadodara, Ahmedabad, Rajkot and Bhuj daily broadcast the prices of important agricultural commodities in the evening. The black boards at the yard display the daily prices. A few selected market committees also supply daily and weekly prices of specified commodities to the Directorate of Economics and Statistics, Government of India. The wholesale prices are published in the local newspapers daily.

The statutory powers of the Director of Agricultural Marketing are vested in the Director of Cooperative Department. The Registrar of cooperative societies, Gujarat state is the head of the department. At the head office level, the Director, Dy. Director, and other subordinate staff of the branch of the regulated market in the registrar's office of cooperative department, Gujarat State, Gandhinagar is responsible for the implementation, execution and superintendence of the regulation of markets. At the district level, the respective district registrars and assistant registrars act as deputy or assistant directors.

The high powered committee in August, 1983 recommend to set up a statutory board. The government of Gujarat established Gujarat State Agricultural Marketing Board whose present Chairman is Shri Shankarlal Guru. The Board has 18 members The chief objective of the Gujarat State Agricultural Marketing Board is to ensure planning and development of agricultural produce markets and to provide technical, financial and legal help to the market committees.

The Board helps the committees to get loans from commercial or cooperative banks under NABARD refinance scheme.

Prior to the existence of the Board, there were 24 talukas where Market Act was not applicable. After the establishment of the Board, the whole state of Gujarat is brought under regulation.

Conclusion:

In this chapter, meaning, definitions and objectives of agricultural marketing in general and regulated marketing in particular have been discussed. Further, the historical development of regulated markets in Gujarat has also been given along with the provisions of the Act.

NOTES

1. R.L. Kohls, **Marketing of Agricultural Products** New York: Macmillan, 1961, p.6.
2. H.S. Irwin, "Intangible Side of Agricultural Marketing - A Neglected Area of Research", **Journal of Farm Economics,** 14 (3) 1973, p.808.
3. James Rhodes V. **The Agricultural Marketing System** Columbia: Grid Publishing, Inc. C.1978, Tomek G. William and Kenneth L. Robinson; **Agricultural Product Prices** London: Cornel University Press, 1972, p.222; Dahl C. Dale and Hammond W. Jorome, **Market and Price Analysis: The Agricultural Industries** New York: McGraw Hill 1977. p.129.
4. A.P. Gupta, **Marketing of Agricultural Produce in India** Bombay: Vora & Co., 1975.
5. E.A.J. Johnson, **Market Towns and Spatial Development in India** cited by Barbara Harriss in "Coarse Grains, Coarse Interventions Regulated Markets". Part I A Case Study of Andhra Pradesh and Karnataka" Working Paper, 2 (Madras Institute of Development Studies, May 1980), p.1.
6. K. Subbarao, "**Presidential address**", Indian Journal of Agricultural Marketing 2(1), June 1988.
7. Frederick, Lundy Thomson **Agricultural Marketing,** New York: McGraw Hill 1951, p.175.
8. C.T. Wong, "Marketing of Locally Produced Vegetables in Hong Kong", **Some Studies of Fresh Fruit and Vegetable Marketing in Asia,** ed. Herman Southworth New York: Agricultural Development Council, 1974, p.1.
9. Government of Gujarat, **Gujarat Development Statistics 1980,** Centre for Monitoring Gujarat Economy, Gandhinagar: Bureau of Economics and Statistics, 1981, p.248.
10. **Regulation of Agricultural Produce Market in Gujarat State,** Gandhinagar: Gujarat State Agricultural Marketing Board, 1988. p.2

II

STUDIES ON AGRICULTURAL MARKETING IN GUJARAT:
A Review

In this chapter, an attempt has been made to review the research studies pertaining to agricultural marketing in Gujarat.

Ranade C.G., Rao, K.H. and Shah, D.C. [1] examine the possibility of increasing incomes of groundnut by doing analysis first at macro level and then at microlevel. From the analysis at macro level, the authors conclude that inspite of wide fluctuations in groundnut production, the private trade and oil industry managed to maintain their profits. During the years (1963-64 to 1977-78), the production declined and the farm harvest prices as well as oil and oil cake prices increased, though the increase in latter prices was more than that in former ones. This was the case at All India level as well as for Andhra Pradesh, Gujarat, Maharashtra and Tamil Nadu. The analysis at micro level is done for a sample of farmers in two major groundnut growing districts - Kurnool and Anantapur in Andhra Pradesh. The micro level analysis shows that the share of groundnut cooperatives in the total groundnut market is insignificant. The only positive aspect of groundnut cooperatives in Andhra Pradesh is that their deductions for weighment and other services are substantially lower than those practised by private traders. There is no vertical integration in the groundnut cooperative in Andhra Pradesh. The groundnut cooperatives in Andhra Pradesh do not follow the pooling system which is followed in Gujarat. The authors have estimated the net benefits of integrated cooperatives to farmers. The increase in the income of farmers after integration will be about 35 to 114 per cent depending upon the marketing channel where vertical integration takes place and it is quite significant.

Agro-economic research centre[2], in its study showed that farmers who sold in the village received less price as compared to the price in the town. Another point to be noted was that the quantity sold per farmer who sold in town market was more than quantity sold per farmer in the village. The storage cost was found more in Kheda because of relatively more humid climate. The survey has shown that the sale of auction method results into most competitive price for the farmer. Auction was not prevalent in Nadiad, Bayad but it was partly practiced in Mehmedabad, Kathial and Talod markets. Again, Mehmedabad, Kathial, Talod and Bayad markets did not have the institution of commission agency. In these markets, except Bayad, auction was carried on. A major reason for not selling in the market yard by the farmers was that it involved considerable time and that they had to wait for long to sell until their turn for auction arrived. For this reason, a farmer may prefer to deliver the goods directly to the trader even at a slightly lower price. It is found that the regulation of marketing practices in many markets is not fully implemented, particularly, in the markets of Kheda and in Sabarkantha. Therefore, it is suggested that the sale by auction should be implemented except for petty sales. The practice of charging unusually high commission and muddat charges in some markets of Kheda should be stopped. In short, prices should be supported by government action in order to assure farmer minimum price.

Charan, A.S., Seetharaman, S.P. and Bapna, S.L.[3] found that out of 184 talukas, 143 talukas in Gujarat had market yards which were brought under regulation by the government. The arrivals in regulated markets were found to be increasing over the years. Regulated markets are more in number in south, middle and north Gujarat and relatively few in Saurashtra. Among many commodities only a few are brought under regulation. In regulated market a large number of institutions of different types function. In case of foodcrops like wheat, bajra and jowar where limited processing is required, private trade is very active. Many market committees over a period of time have not gone beyond performing the function of regulation. The value addition tasks like arranging transportation, grading, providing price information at different markets, developing better packaging practices etc. have been neglected. The total number of traders and commission agents registered with marketing committees in 1977-78 was 109880 which worked out to an average of about 300 intermediaries per regulated market. Gujarat has the largest number of marketing and processing cooperatives which function effectively. The study made a few suggestions and one of them is that the function of regulated markets needs to be completely revamped. Regulated markets, as they function at present, provide only physical and regulatory facilities. Real benefits to the farmers would accrue only when the price determination process is strengthened and facilities like transportation, storage, grading, packaging, price information and institutional financing are developed.

Gupta, V.K. and Mathur, D.P.[4] hypothesised that favourable environmental and economic factors would reveal changes in production and marketing behaviour of farmers. This is a comparative study of Punjab and Mehsana. This study has revealed a decline in the level of closing stocks maintained by farmers during 1974-75 and 1975-76 as compared to what it was in 1973-74. A decline in absolute sense was observed in the marketed surplus among large farmers in Mehsana in 1975-76 as compared to what it was in 1974-75. Large farmers in Mehsana tried to make their loss making enterprise profitable by holding stocks in expectation of getting better prices. In Mehsana, farmers disposed off their produce in a greater measure in 1975-76 at village level as compared to what they did in 1973-74. Most of the sales at village level were to the resident agents and small traders who had direct links with wholesalers and retailers in urban areas. The policy of farm levy during periods of bountiful supplies and depressed open market prices is against the farmers interest in areas like Gujarat.

Patel, S.K.[5] attempted to study the important problems that the tobacco growers face in marketing their produce in Kheda district of Gujarat state. These problems include high rates of deductions both in weight and payment, low prices of tobacco, delay in payment, lack of proper storage facilities, fluctuations in tobacco prices, lack of easy and adequate finance, absence of established marketing centres, standard code of transactions and lack of scientific grading. It is found that the farmers hardly get the due share of their produce. It is also true that so far adequate attention has not been paid to the regulation of tobacco industry to protect the interest of tobacco growing farmers.

Vyas, V.S. and Maharaja, M.H.[6] make a few broad conclusions of the study. Total marketable surplus as a percentage to total crop production was generally found to be high in the higher size groups. The difference between the share of marketable surplus in the lowest size group and the highest size group was fairly marked. It is found that the marketable surplus of commercial crops was determined almost exclusively by the production of commercial as well as superior cereals. The overall marketable surplus was in the range of 93-94 per cent of the total production. However, the requirements for family consumption and other disposals also exercised statistically significant influence. The marginal increase in marketable surplus in relation to increase in production of superior cereal as well as coarse cereal was positive and significant.

Acharya, S.S.[7] states that the arrival of pulse crops in markets have fluctuated widely from year to year. Nearly two third of pulse arrivals of Gujarat State are from Panchmahal district. Dahod market alone accounts for more than 50 per cent of arrivals of gram in Panchmahal. In Dahod, 51 per cent of the total arrival are from Madhya Pradesh and 25 per cent from Rajasthan. Price policy adopted by local traders/processors attracts pulse production from these neighbouring states. A very clear seasonality in market arrivals was observed. Nearly 66 per cent of gram arrived between February to June, 78 per cent of arhar arrived between December to March and 70 per cent of urad arrived between October to December. There were two peaks in mung arrival. One in September-October (38 per cent) and two in May-June (44 per cent). Secondary arrivals account for more than 86 per cent of total arrivals in Dahod market. On an average, 40.2 per cent of pulse grains arriving in Dahod market were dispatched to outstations. Stocks held by traders were closely associated with the pattern of arrival, i.e. stocks were more when arrivals were more and vice versa. Nearly two third of quantity was sold to cooperative societies and remaining one third in the regulated market. Nearly 27 per cent of the farmers sold gram in the lots of 2.1 to 4 quintals and 19 per cent sold in the lots of more than 20 quintals. Bullock cart was the main means of transportation both for gram and arhar. Nearly three fourth of gram and half of arhar was carried in heaps and remaining in bags. Delay in payment was not a serious problem with the farmers who sold their produce. The marketable surplus was 81 per cent in case of gram and 74 per cent in case of arhar. There was no significant difference in marketable and marketed surplus. On farm consumption was 10.5 per cent of gram and 21.2 per cent of arhar. Average marketed surplus per farm was 12.25 quintals of gram and 2.64 quintals arhar. The marketing cost incurred by the farmer was negligible but it was more when the produce was sold in the regulated market than to the cooperative society mainly because of transportation cost. Average cost of marketing was Rs.1.17 per quintal for gram and Rs.3.78 per quintal for arhar. Farmers received higher price by selling in the regulated market than to the cooperative society. Price difference was more than the difference in marketing cost for sales at these two places.

Bapna, S.L. and Rao, K.R.[8] developed a technique for predicting output, market arrival and prices by using the available information on prices, weather and technological variables in the pre-harvest period. Such information is very crucial for decisions to be made on inventory management and for developing a suitable marketing policy on procurements, distribution, support prices, zoning, import-export and a host of related decisions by the government, industrial firms using raw-materials from agriculture, consumers and others concerned with optimal allocation of scarce resources. The variables predicted are output, market arrivals and prices. The approach used to identify the factors affecting these variables was to first carefully examine the variables by a graphic as well as tabular analysis and then develop supply and price models in which alternative specifications of the variables were attempted. It is shown that the information required to make a prediction about these variables is available before harvest and by incorporating these in selected equations, reliable outlook information can be generated. The study made predictions of bajra and groundnut output by estimating supply function for output and its components namely area and yield. Supply functions were fitted on time series of twenty five years, starting from 1957-58 and ending in 1981-82. Predictions were provided for the year 1982-83, 1983-84 and 1984-85. Output and yield supply functions for bajra and groundnut had very high explanatory power. R^2 varied between 0.80 to 0.90 and had low standard errors. The response of area under bajra and groundnut to respective prices was generally positive and statistically significant. However, the elasticity coefficient in bajra area functions were higher than groundnut functions particularly in Saurashtra where groundnut was a dominant crop. Two types of market arrivals and prices in Gujarat were predicted. There were total market arrivals and post harvest market arrivals and annual wholesale prices and post harvest prices. For predicting market arrivals and prices, functions were estimated for eighteen years starting from 1964-65 to 1980-81 and predictions were provided for 1981-82, 1982-83 and 1983-84. Market arrivals and prices are simultaneously determined in the market. Therefore, in market arrival equations, estimated prices were used by which the influence of market arrivals was removed. The study had certain limitations. The study used disaggregate data only for the broad political regions of Saurashtra and Gujarat. However, more homogeneous regions need to be formed in order to obtain precise estimates. Rainfall variable was defined on the calendar month basis. However, the crop growth pattern does not follow the calendar month schedule. If efforts to estimate the outlook in different states are made, the assumptions on production in other states would be more reasonable. The above limitations according to the authors could be removed but it requires major research work together with adequate resources and considerable interdisciplinary work.

Thakur, D.S., Chauhan, S.K., Sharma, K.D.[9] attempt to give in their paper a brief history, growth and coverage and to study the functioning of regulated markets in Gujarat and Himachal Pradesh. They used two criteria for analysing the economic or pricing efficiency viz. market integration and price spread. The degree of market integration is estimated by working out correlation coefficients between wholesale prices in the markets. It was found that coefficients for prices of bajra are high only for Borsad, Harij, Mehsana and Unjha and for other markets, it is low and even negative. The analysis of market margins and price spread for those commodities show that the net price given to farmers ranges from 68 to 75 per cent of the consumers' price and the rest goes for marketing margins and costs. The largest slice of the marketing cost is taken by traders as their commission or profit which account for 70 to 80 per cent of marketing cost which is again a clear sign of marketing inefficiency. Similar conclusion emerge from the analysis of data with respect to Himachal Pradesh.

Madalia, V.K.[10] highlights the details of the functioning of regulated market in Surat. The paper shows that the total benefits accrued to the farmers during 1985-86 were of the order of Rs.1.90 crores and the cumulative benefits since 1951 were of the order of 14.51 crores. The farmer gets substantial benefit as the market charges are paid by the buyers. The author makes suggestions like clearance of land acquisitions cases by the government, providing training to the supervisory staff, creation of cold storage facility and provision of transportation facility and starting of a journal on marketing. The cooperatives should involve themselves as commission agents and traders. Marketing of fruits and vegetables should have separate acts.

Bhatt, B.D., Antaniasd, K.L., Shiyani, R.L.[11] have collected monthly and annual market arrival and prices for nine years i.e. 1978-79 to 1986-87 from Ahmedabad Regulated Market. The linear growth rates worked out from the annual data revealed that the prices of potato and onion as well as arrival of chillies increased significantly. The authors observed that the seasonal and cyclical variations of prices were higher than the variations of arrivals in all the crops except chillies. The irregular fluctuations were higher in respect of arrival as compared to the prices of all the crops except brinjal. The coefficients of variation showed that the arrival and prices of all the crops were influenced more due to cyclical variations than rest of the components and the prices of chillies were influenced comparatively more due to trend variation.

Brahmbhatt, D.M.[12] studied mango marketing in Valsad district with years 1979-80 and 1980-81 as reference period. Three talukas viz. Valsad, Gandevi and Umargaon of Valsad district were selected to make cross section study of mango marketing problems in the region. From three talukas, eight villages were selected keeping in view their nearness to mango market and location from pucca road. From selected villages, a sample of 120 orchardist households was drawn by stratified random sampling. It was found that there was a tendency of the positive correlationship between the proportion of operational holding devoted to the crop and the orchard size group. Mangoes like other fruits are sold in three ways - open auction, under-cover and private negotiation. Mangoes, being perishable, cannot be stored for long and hence supplies fluctuate every day giving rise to wide price fluctuations. The demand for mangoes is widespread but its supply is localised. Inadequacy or inefficiency in transport in terms of either limited directions or up to certain distances give rise to glutting of mangoes which in turn leads to quality deterioration and hence lowering of prices. Cooperative marketing societies for Gandevi taluka, commission agents and traders in market yard for Valsad taluka and terminal agents for Umargam taluka were the important agencies for mango sale in the district. Cooperative societies of Gandevi taluka act as an agent of the orchardist member by sharing him three per cent commission on the gross sale value of mangoes. The society charges one per cent on the value of mangoes pooled from the members before selling the produce to the trader. The author makes a few suggestions: (1) The government should arrange for regular broadcasting of price information through its radio station during mango marketing season. (2) More regulated agriculture market yards should be opened. (3) Research should be done to evolve late maturing hybrid Alphonso variety and which has dwarf size trees instead of existing tall trees. (4) Old trees need to be replaced by new and vigorous trees to achieve better yield levels. (5) The agriculture extension agency should endeavour to bridge communication gap between research laboratory and mango orchard. (6) There is a need to standardise unit of price quotation in all mango markets. (7) Efforts should be made to extend cold storage facilities.

Joshi, V.R.[13] made an attempt to study the working of Regulated Markets in the whole state of Gujarat. The concern of the study was with the establishment and growth of Regulated Markets in the state along with the legal framework, their functions and financial position. The study brought to light the extent of sales outside Regulated Market, the area of unregulation, the ultravires transaction taking place within the limits of the market proper, the wide disparities in the bases and rates of market fee etc.

Raju, V.T., and Kakadia, B.H.[14] revealed that the farmers of Rajkot District have sold most of their production of groundnut making provision mainly for seed. The percentage of sales of groundnut to total production increased with the increase in the size of holding. The regulated markets were found to be the important marketing channels followed by cooperative and village merchants for groundnut marketing. The price of groundnut in Gondal market and Rajkot market showed an increasing trend. The correlation coefficiency between monthly prices and arrivals was found negative in both the markets indicating a definite inverse relationship i.e. groundnut arrival increases, price decreases and vice versa.

Pandey, M.K.[15], found that large farmers formed nearly sixty-three per cent of the sample and contributed to more than two-third of paddy production and three-fourths of wheat and bajra production. The crop loans are disbursed on a large scale by cooperative societies with a view to helping the farmers in increasing their agricultural production. The foodgrain market was found having conditions of nearly perfect competition in Ahmedabad district. The author makes a few suggestions: (1) The Gujarat State Cooperative Marketing Federation should have its own milling units with adequate capacities. (2) The foodgrains collected by the village cooperatives may be handled by taluka/district Purchase and Sales Unions before they reach the Federation. The cooperatives can establish processing units for production of rice flakes, wheat flakes and fine wheat flour. The author further suggests that the foodgrain marketing can be improved considerably by establishment of regulated markets at secondary level and by proper integration of cooperative marketing societies with them.

Thakur, D.S.[16] analyses the operational efficiency of the foodgrain marketing system and secondly evaluates the pricing efficiency of the foodgrain markets. The author finds that the sale direct to the consumer fetches the highest net price to the farmers. The sale through retailer is the next possible profitable channel for the farmers. The sale through wholesalers and commission agents is the third best channel and much more remunerative as compared to the sale through the cooperatives and village sales through the village and the itinerant merchants. The net income of the farmer increases in the order of sale through the village merchants and cooperatives, sale through itinerant merchant and millers, sale through the agent middlemen, sale through the retailer and sale direct to the consumers. The analysis of pricing efficiency in the study shows that the existing foodgrain marketing system on the whole is not efficient. There are only a few traders in most of the markets who purchase most of the produce regularly in the absence of pure competition. They generally agree through mutual understanding and outright collusion to avoid price competition. The overall analysis of the marketing margins and price spread in the foodgrain marketing system thus shows that even in normal years the traders' profit margins account for quite a large proportion of the price paid by the consumers which is clearly a sign of pricing inefficiency. According to the author, the farmers want strict supervision against pilferages, safeguarding the produce against losses through damage, reasonable price for their produce and quick cash payments etc., all of which have a bearing upon the prevailing shortcomings and inefficiencies of the existing marketing system.

Ranade, C.G., Singh, R.B., Rao, Hanumantha[17] in their study examine the price spread in cooperative and private marketing channels in Gujarat. According to the authors, if cotton production is to be encouraged in India, and if economic status of cotton growers is to be enhanced, the cotton growers should get a fair price for the produce they sell. A hypothesis behind this study is that the price received by farmers depends upon the channel through which they sell kapas. They have selected two major cotton growing districts in Gujarat, viz. Surendranagar and Sabarkantha. In Surendranagar, the authors find four distinct types of private channels. Out of these four, only one is under regulated market system. The benefit of the regulated market is fetched by the traders and not by the farmer. Another interesting finding is that in all the channels under private trade the small farmer is receiving lower price than that received by the large and medium farmers. In cooperative channel, large as well as small farmers receive an equal share in the price of lint and seeds. Among intermediaries, it is the small trader such as village trader who bears higher risk of marketing than ginners and wholesale traders. In cooperatives, it is the farmer who bears the risk of marketing. The study shows that the share of farmers in the price of lint and seeds is lower in an area dominated by private trade than in private channel where cooperatives also function simultaneously. The survey of farmers reflect that the farmers do require much higher marketing advance when they deliver kapas to cooperatives than currently offered by cooperative credit institutions. It was found that economic variables do have an influence over the acreage allocation decisions. Cooperative marketing is found much more beneficial to small farmers than the private trade channel. The study suggests that in areas where private trade is dominant and the benefits of regulated market do not percolate to farmers, government should intervene, so that farmers can sell kapas in regulated market.

Jasdanwalla, Z.V.[18] worked on an extensive field work and elaborate statistical data. The major findings of her study are: (1) The main requirements for efficient and perfectly competitive market conditions are substantially satisfied in the Rajkot district. (2) For groundnut, analysis showed the market to be efficient and well integrated at almost all stages between the producer and the consumer. (3) Production was found to be responsive to market forces. (4) In the last decade relative price movements have been in favour of cash crops as opposed to foodgrains.

Conclusions

Studies pertaining to agricultural marketing in Gujarat referred above cover the following aspects:

(1) The studies are crop specific and area specific.
(2) They throw light on trends in market arrivals and price fluctuations in terms of seasonal and cyclical variations.
(3) They focus attention on price spreads and marketing margins.
(4) They discuss benefit of alternative marketing channels to the farmers.
(5) They describe composition of market committees, Market Regulation Act and the rules and regulations of regulated markets.
(6) They examine the impact of marketing on cropping pattern and marketable surplus.

Like other market studies, the present study does discuss market arrivals and prices but it has been done so as to examine the various aspects of marketing like competitiveness of market, market integration, market stability etc. The present study distinctly differs from other studies in the sense that it discusses the problem of marketing from the point of view of farmers belonging to different size groups viz. marginal, small and large farmers. Thus the present study has a definite objective in view. The objective is to examine the efficiency of regulated markets in terms of competitiveness, market integration, market stability, and benefits to marginal, small and large farmers inside regulated market yards and outside regulated market yards. This is how the present study differs from the earlier studies made on agricultural marketing in Gujarat, the details of which are given in the next chapter.

NOTES

1. C.G. Ranade, K.H. Rao and D.C. Shah, **Groundnut Marketing : A Study of Cooperative and Private Trade**, CMA Monograph No. 92, Ahmedabad : Indian Institute of Management, 1982.
2. Agro-Economic Research Centre, **A Study of Production and Marketing of Bajra in Three Districts of Gujarat** (Banaskantha, Sabarkantha, Kheda), Vallabh Vidyanagar, Sardar Patel University, 1972.
3. A.S. Charan, S.P. Seetharaman, S.L. Bapna, "Agricultural Marketing System in Gujarat : A Perspective", A paper presented at Fifteenth Gujarat Economic Conference, Surat, October-November 1983.
4. V.K. Gupta and D.P. Mathur, **Changing Pattern of Wheat Production and Marketing**, CMA Monograph No. 67, Ahmedabad : Indian Institute of Management, 1977.

5. S.K. Patel, "Problems of Tobacco Marketing in Kheda District", A paper presented at the Fifteenth Gujarat Economic Conference, Surat, October-November, 1983.
6. V.S. Vyas and M.H. Maharaja, "Factors Affecting Marketable Surplus and Marketed Surplus : A study in Two Regions of Gujarat and Rajasthan", **Arth Vikas**, 2(1) January 1966, pp.52-78.
7. S.S. Acharya, **Agricultural Production, Marketing and Price Policy in India**, (A Study of Pulses) New Delhi : Mittal Publishers, 1988.
8. S.L. Bapna and K.R. Rao, **Supply and Price Outlook for Crops : A study based on Pre-harvest Market Information in Gujarat**, New Delhi : Oxford and IBH, 1987.
9. D.S. Thakur, S.K. Chauhan, K.D. Sharma, "Efficiency and Weakness of Regulated Markets", **Indian Journal of Agricultural Marketing**, 2(1), June 1988.
10. V.K. Madalia, "Functioning of Regulated Markets at Surat and Its Impact," **Indian Journal of Agricultural Marketing**, 2(1), June 1988.
11. B.D. Bhatt, K.L. Antaniasd, R.L. Shiyani : "An Analysis of Arrivals and Prices of Important Vegetable Crops in Ahmedabad Regulated Market in Gujarat State," **Indian Journal of Agricultural Marketing**, 1(1), June 1988. p73
12. D.M. Brahmbhatt, "Mango Marketing in Valsad District", Paper presented at the Fifteenth Gujarat Economic Conference, Surat, October - November, 1983.
13. V.R. Joshi, **Regulated Markets in Gujarat**, Vallabh Vidyanagar : Sardar Patel University, Gujarat, 1971.
14. V.T. Raju and B.H. Kakadia, "Marketing of Groundnut in Rajkot District of Gujarat State", **Indian Journal of Agricultural Economics**, Volume 3, July-September, 1984.
15. M.K. Pandey, "Marketing Efficiency of Foodgrain in Gujarat" (Unpublished Ph.D.Thesis), Gujarat University, 1982.
16. D.S. Thakur, "Foodgrain Marketing Efficiency : A Case Study of Gujarat", **Indian Journal of Agricultural Economics**, October-December, 29(4), 1974.
17. C.G. Ranade, R.B. Singh, Hanumantha Rao, **Marketing Channels and Price Spread in Cotton**, CMA Monograph No.83, Ahmedabad : Indian Institute of Management, 1979.
18. Z.Y. Jasdanwalla, **Marketing Efficiency in Indian Agriculture**, Bombay : Allied Publishers, 1966.

III

DESIGN OF THE STUDY

This chapter proposes to select the study area, markets, crops, study period, approach to the study, performance variables, tools of analysis and hypotheses.

Selection of the Study Area

Eventhough the purpose of the present study is to understand the working of agricultural marketing in Gujarat, the attention is focused on a detailed study of the district. For this purpose, the selection of the district becomes imperative. The norms for the selection of a district include the level of agricultural prosperity,of the district and the number of Agricultural Produce Market Committees and the number of market yards. From the point of view of the level of agricultural prosperity, Valsad district stands first but from the point of view of the number of Agricultural Produce Market Committees and market yards Valsad ranks very low. On the other hand, though Mehsana has a lower rank from the point of view of the level of agricultural prosperity, it stands second to none in terms of the number of Agricultural Produce Market Committees and market yards. Moreover, the phenomenal growth of Unjha market yard is unique and unparallel. The factors affecting the extraordinary growth of Unjha market needs special study and a study has been already undertaken on this.[1] The growth of Unjha market cannot perhaps be a representative of the growth of other markets of Gujarat state. Moreover, to study regulated markets of Mehsana district excluding Unjha market will also distort the picture of the growth and development of regulated markets in Mehsana District. For this particular reason, Mehsana district is not selected for the present study. Looking to the above difficulty, Kheda district was found appropriate for the purpose of the study because on one hand it is fairly developed in terms of agricultural prosperity and ranks second in terms of the number of Agricultural Produce Market Committees and market yards (see Table 1.1).

Selection of the Study Period

In Kheda district the first Agricultural Produce Market Committee that started functioning was that of Kapadvanj in the year 1947. The last Agricultural Market Product Committee was established in the year 1960 in Petlad. During this period of thirteen years, as shown in Table 3.1, eleven market committees were established in the district of Kheda. However, Green Revolution came into being in 1965 and moreover the newly established markets might be in their infancy. On the other hand, Gujarat State experienced severe drought after 1985 and this might have affected the agricultural product and the markets arrivals. For these reasons, the study was restricted to a decade beginning from 1975-76 to 1984-85. This can be considered the most ideal period to study the working of regulated markets of Kheda district.

TABLE 3.1: Growth of Regulated Markets in Kheda District

Year	Number
1947	1
1948	5
1950	6
1955	8
1956	9
1958	10
1960	11

Selection of Crops

The present study is restricted to three crops viz. paddy, wheat and bajra. These crops were selected on the basis of area under different crops as a percentage to total cropped area of the district. The area under paddy, bajra and wheat as a percentage to total cropped area of the district are 29.57, 27.55 and 11.27 respectively in the year 1984-85. Though tobacco has a relatively high percentage area under the crop (17.78), it is not selected for the purpose of the study because it is not a regulated crop and further its marketing is organised at the state level.

As such crops selected for study are three in number, so far as paddy is concerned the arrival and price of paddy and rice are given separately in the annual reports of market committees and therefore they have been treated separately in the present study if they are separate crops.

Selection of the Markets

There are eleven Agricultural Produce Market Committees in Kheda district. It is not possible to study in much detail all the regulated markets. Therefore, it was decided to select four markets of the district for the purpose of the study.

Ideally speaking, markets should be selected on the basis of markets arrivals as a percentage of marketable surplus of the area. However, the difficulty is that the figures pertaining to marketable surplus of different crops in different market areas were not available at the time of selection of markets. Therefore, the resort has been taken to the market arrivals as a percentage of the production of different crops in the area. Again, at the taluka level (where a market is established) figures pertaining to area under different crops are available but the yield per hectare estimated on a scientific basis at the taluka level is not available.

Looking to these difficulties, figures pertaining to production of different crops in different market areas (talukas) were obtained from the unpublished records of Directorate of Agriculture Marketing, Gandhinagar and then market arrivals of different crops as a percentage to the production of the respective crop in the area was worked out. This has been shown in Table 3.2.

At this stage, one more difficulty was experienced by the author in the selection of markets. The purpose of the study is to understand the working of marketing mechanism in the selected three markets with respect to the selected three crops. The selected three common crops did not rank high in terms of arrival as a percentage of production in the selected four markets, e.g. in Matar regulated market, arrival of paddy as a percentage of production of the area is very high (71.9) but for other crops like bajra, its arrival is low (7.48). Therefore, Matar could not be selected as the study intends to study all the three crops in all the four markets. Thus, markets which showed high arrivals of only one crop have not been selected. In the same way, though the performance of Nadiad and Anand regulated markets was good in terms of arrival as a percentage of production, they were not selected

TABLE 3.2: Production and Market Arrivals of Paddy, Wheat and Bajra For the Year 1984-85 in the Regulated Markets of Kheda district

(In quintals)

Sl No.	Name of regulated market	Paddy		(3b) as %	Wheat		(4b) as %	Bajra		(5b) as %
		Production	Arrivals		Production	Arrivals		Production	Arrivals	
(1)	(2)	(3a)	(3b)	(3a)	(4a)	(4b)	(4a)	(5a)	(5b)	(5a)
1.	Nadiad	401920	180541	45.92	66980	102262	152.67	280220	48037	17.14
2.	Anand	265050	26505	10.00	51840	39879	76.93	124870	57429	45.99
3.	Thasra	178380	-	-	33490	-	-	84590	-	-
4.	Borsad	191300	186817	97.65	36680	48530	132.30	299070	81087	27.11
5.	Petlad	540430	19362	3.52	57530	5036	8.75	250930	6509	2.59
6.	Khambhat	742310	169522	22.83	89870	30996	34.48	251880	36959	14.67
7.	Matar	724090	520625	71.90	113860	45040	39.56	112930	8444	7.48
8.	Mehmedabad	212170	286964	135.25	82990	84844	102.23	129330	39604	30.62
9.	Kapadvanj	107090	73251	68.40	118210	29971	25.35	348710	31305	8.97
10.	Balasinor	90830	-	-	35700	-	-	156260	-	-

Source Directorate of Agricultural Marketing, Gandhinagar.

due to certain administrative difficulties and practical considerations. To have reasonable arrivals of the selected three crops in all the four markets, the markets selected are Borsad, Khambhat, Kapadvanj and Mehmedabad.

Choice of Approach

The approach selected for the present study is known as "Market Structure Approach". The market structure approach has been explained by authorities like Purcell, Bain and Barbara-Harria. Their explanations are like this:

> Purcell[2] defines the market structure as the organisational characteristics of the market system. According to him, the principal characteristics of market structure are seller concentration, buyer concentration, barriers to entry and the degree of product differentiation.

According to Bain[3], the market structure refers to the organisational characteristics which determine the relations of sellers in the market to each other, of buyers in the market to each other, of the sellers to the buyers and of sellers established in the market to other actual or potential new firms which might enter the market. To Barbara Harriss,[4] the market structure is the organisational characteristics influencing entry, competition and price formation in the market.

This approach suffers from certain limitations and has produced contradictory results in the empirical studies conducted elsewhere.[5] In certain other studies, this approach has produced revealing results.[6] Therefore, this approach has been selected for the present study. Further, the organisation has been studied keeping in view the main economic aspects of market study such as efficiency, integration,stability, competitiveness etc. Non-economic aspects such as establishment and administration have been ignored here in this study. Certain aspects of the study of marketing such as the degree of production differentiation (Purcell) relation of sellers to potential new firms (Bain) and price formation (Barbara Harris) have not been studied here for practical difficulties.

Economic aspects of the market study have been studied here keeping in view the perfect market as discussed in text-books of micro economic theory. This is because micro-economic theory considers the perfect market as the most efficient and therefore an ideal one. Instead of taking all the conditions of a perfectly competitive market, only few selected conditions of a perfect market have been taken here, following Bressler and Richard[7] because they consider these conditions sufficient for the purpose of studying the efficiency of Agricultural Marketing.

(1) perfect knowledge by all buyers and sellers
(2) each buyer and seller acts in an economically 'rational' way disregarding any influence of his actions on price, and
(3) free entry in all directions.

If these three conditions are fulfilled, a market becomes impersonal and competitive. As regulated markets come into existence through legal procedure with a definite objective of protecting the interests of the farmer, some additional conditions become necessary to judge the performance of regulated markets.

(i) market openness
(ii) absence of excessive profits to transactors
(iii) immediate payment
(iv) absence of delay in transaction
(v) absence of corruption
(vi) controlled oscillation of market prices around normal prices
(vii) amenities to market participants
(viii) helpfulness of the market staff and functionaries.

Thus the total conditions turn out to be eleven to judge the performance of a market. A market which satisfies these eleven conditions will be considered as an ideal market for the purpose of our analysis. The deviation of the working of an actual market from an ideal one has been taken as a measure of inefficiency of the regulated market.[8]

Regulated market does not mean that market mechanism is eliminated. It simply means that unwanted predatory middle-men are eliminated and the exploitation of the farmer is eradicated. Though the regulated market functions within the legal framework, it still resembles the private market because market forces of supply and demand have a free play. The only difference between the regulated market and the so-called free private market is that in previous one, the trade practices are regulated by law and hence they are subject to certain constraints. Therefore, the evaluation of regulated markets should be done in the light of these constraints which have been highlighted by the experts[9] in the field of Agricultural Marketing. Lele is of the opinion that regulated markets are established with multiple economic and social objectives. These objectives include (i) protection of the interest of the producer-seller through healthy competition and ensuring a fair deal; (ii) improvement of marketing services; (iii) equalization of the bargaining power of the farmer and trader, (iv) creation of greater competition and better services of transport, storage, grading, pooling, market intelligence, (v) inspection of weights and measures (vi) provision of credit, finance etc.

The experts of the Agricultural Development Council have elaborated the difficulty in evaluating the performance of market. According to them, "The question of what is good marketing system cannot be separated from the more fundamental question of what is a good society, for the evaluation of a market organisation has meaning only within the context of a broader view of the 'good society' or 'good life'". Thus the performance of regulated markets should be judged in the light of given objectives and the actual achievements. However, performance of regulated market cannot be judged only through goals i.e. economic and social objectives lying behind the law, because the constraints in terms of economic and social objectives have a bearing on their working. The objectives of market regulation do not exactly coincide with the conditions of perfectly competitive market, yet regulated market has much similarity with a perfectly competitive market.

Thus the above discussion shows that the performance of a regulated market must be judged with the help of a dual criteria viz. (i) fulfilment of the goals and objectives, and (ii) satisfaction of the conditions of a perfectly competitive market.

Selection of Performance Variables

To judge the working of a regulated market with the help of dual criteria viz. (i) fulfilment of goals and objectives, (ii) satisfaction of some of the conditions of a competitive market, two types of variables are selected in the present study - (a) economic variables, and (b) non-economic variables. This dichotomy need not be stretched too far while actually analysing the problem. Economic variables include (i) association of prices (2) parity prices (3) marketing margins (4) marketing procurement costs and (5) market arrivals. Non-economic variables include (1) disposal quickness (2) immediate payment (3) amenities to market participants (4) helpfulness of the market staff and functionaries to transactors (5) corruption and crime (6) pilferages and wastes (7) market consciousness and (8) market information.

From the economic variables mentioned above, "producer's share in consumer's rupee" as one of the aspects of marketing margins have been excluded in this revised version of the study because of the unreliability of retail prices required to work out the marketing margins. In case of parity price, though the exercise has been carried out, it has been given in Appendix I for the reason that it does not fit within the structure of the study.

Economic Variables

(1) *Association of Prices*

Price stability is one of the essential characteristics of efficient market[10]. If prices are not stable, it creates uncertainty on the part of farmer/producer. In subsistence agriculture, where most of the farmers are small and marginal farmers, unstable prices i.e. price-uncertainty adversely affects the economic conditions of a poor farmer[11]. Moreover, there is a great deal of disincentive to invest in traditional agriculture because of price fluctuations over space and time[12].

Further, prices are more variable in price discovery system than in price setting system which is true for regulated markets.[13] Therefore, price stability becomes more essential for a regulated market.

So far as prices are concerned, another important variable is that markets should be integrated. In other words, price differences over space should not exceed transport cost. However, some experts have raised objections against using price variability as an indicator of market inefficiency because price is determined by forces which lie outside the sphere of marketing.[14] The implicit assumption behind integrated markets is that buyers and sellers are rational in their behaviour i.e. sellers move from low price market to high price market and buyer move from high price market to the low price market. Market integration has been examined in most of the studies with the help of correlation coefficient between the prices of two markets. High correlation coefficient means markets are integrated and markets are highly competitive. However, in some of the studies this tool has been found as an inadequate one to prove or disprove market integration.[15] Spatial price is found to be more than transport cost even when correlation coefficient is very high[16]. Therefore, parity of prices between two markets should be considered better to judge the market integration.

In the present study, the market integration is examined by correlation coefficient and analysis of variance following Lele[17].

(2) *Marketing Margins*

Marketing margin is generally considered as an index of efficiency. If marketing margins are high, then marketing system is considered an inefficient one and vice versa. Of course, this approach is not free from criticisms. It is argued that a good and honest businessman will charge a higher price and expect higher profit as compared to what is being charged by a dishonest and corrupt businessman. Thus we look at the marketing margins only ignoring the trade practices, they do not give a correct idea about marketing efficiency.[18]

Marketing efficiency can be really judged only when we have data about the marketing charges and traders' buying cost and profits inside and outside the marketyard under the implicit assumption that behaviour pattern on the part of the transactors is ideal and/or identical. In the present study, marketing charges and cost of buying and selling have been estimated inside and outside the yard. This has been done with the help of data collected through primary survey. The cost of selling on the part of the farmer and the cost of buying on the part of the trader are calculated in this study per quintal of the product bought and sold at the time of first sale made inside the regulated market yard and outside regulated market yard. Various charges in regulated markets are fixed by the market committees within the limits fixed by the state government. Information about market charges have been collected from various regulated markets. For outside the regulated marketing system, the information was collected from the farmers and traders through primary survey. Thereafter, the relative efficiency of the two marketing systems was compared through the differences in costs and prices.

(3) *Market Arrivals*

The main objective behind the establishment of regulated market is to protect the interest of the farmers. Protection of traders' interest is neglected in the legislation itself. The reason may be that in developing countries the elected representatives of the state assembly and the ideologies of the political parties to which they belong might be aiming at helping the weaker sections of the society through state legislation. Further, it is normally presumed that the traders' bargaining power is more and they are better informed also. Thus, regulated markets are serving the purpose to the extent farmers use regulated markets to dispose of their produce. In other words, market arrivals can be used as a yardstick to measure the efficiency of regulated markets.

Actually market arrival as a percentage of marketable surplus gives an idea about the efficiency of a regulated market. If this percentage is high, then the market can be considered as an efficient one and vice versa. The difficulty in Gujarat state is that the data pertaining to marketable surplus is not available either from government agencies or from research institutions for a long time span. Occasional studies pertaining to one or two districts are available for Gujarat[19] and they do not serve the purpose of the present study because they are relatively old and cover districts other than Kheda district. Therefore, an attempt has been made in the present study to estimate marketable surplus from a sample study. The details of the sample are given later. The marketable surplus has been estimated with the help of following equation:

$$Am = Sm - Qe - Qc$$

where Am is the maximum possible level of market arrival by regulated markets.

Sm is marketable surplus.

Qe is quantity sold within the exemption limit

Qc is quantity sold as a commitment to clear the loans in kind.

Non-economic Variables

Besides economic variables, non-economic variables are also examined in the present study by interviewing the sample farmers and traders. This is done in the light of the observations made by the investigators during their visits in the field. The non-economic variables that are examined are: grading, disposal quickness, delay in payment, amenities, helpfulness of the market staff and functionaries, corruption and crime, pilferages and wastes, market consciousness, market information etc.

Sampling Procedure and Technique

The sample study was undertaken (1) to estimate marketable surplus of farmers belonging to different size groups, (2) to estimate the cost of selling and cost of buying inside the yard and outside the yard, (3) to find out the reasons for low arrival in regulated markets yards, (4) to find out the prices received and quantities sold inside and outside the market yard, and (5) to know the working of non-economic variables.

In the sampling procedure, it is assumed that for the regulated market to have a larger arrivals, the market should be easily accessible and easily negotiable by both the farmer and the trader. According to the National Commission on Agriculture[20], "to get a fair deal a regulated market must be available within a radius of 5 kilometres, which is negotiable by foot or by cart by the farmer".

Wong[21] is of the view that for perishable commodities, the market should be accessible even by a bicycle track. According to him, no farmer would find renumerative to carry his produce to the market, if it is more than 30 minutes walking distance.

Hence for the present study, villagers within a radius of 5 kilometres of the market area were selected at random. Two villages within a radius of 5 kms and one village within a radius of 15 kms were selected from each market to study the impact of the distance factor on the market arrivals. Thus three villages from each market area were chosen. As four selected markets were to be studied, in all, twelve villages were selected and studied.

For each market 90 farmers were interviewed. Thus in all 360 farmers from four markets areas were interviewed and these farmers were from small, medium and large groups. For studying the traders' responses, 10 traders were chosen from each market randomly. Hence 40 traders were interviewed from four market areas.

Sample Villages

For studying the farmers responses and their marketing costs inside and outside the regulated markets yards, 3 villages from each market area were selected. Out of these three, 2 were within 5 kilometres and one within 15 kilometres of the regulated market yard.

For Borsad market, Davol and Nisariya villages which are within the radius of 5 kilometres of the market and Napa Talpad which is within 15 kilometres were selected for the present primary study. Similarly for Khambhat market area, Shakarpur and Metpur (within 5 kilometres) and Ralej (within 15 kilometres) were selected. For Kapadvanj market area, Nawagam, Jaloya were selected which are within 5 kilometres radius and Atarsumba which is within 15 kilometres was chosen. For Mehmedabad market, Khatraj and Kachhai village (within 5 kilometres) and Modaj village (within 15 kilometres of the market) were chosen.

Data Requirements and Sources

In the present study, both primary as well as secondary data were collected. The details of the primary data are given in the preceding section. The secondary data on prices, arrivals, market fee, licence fee etc. were collected from various sources. They are: (1) State Agricultural Marketing Board, Gandhinagar, (2) Directorate of Agricultural Marketing, Gandhinagar, (3) Published and unpublished records and reports of the Agricultural Produce Market Committees of different talukas of Kheda district.

Questionnaire

A common questionnaire for all sample farmers and a common questionnaire for all sample traders were prepared.

Tools of Analysis

The following tools of analysis are used in the present study:

(1) Classical time series analysis.
(2) Functional Analysis.

(1) *Classical Time Series Analysis*

This analysis is employed to understand the effect of trend, cyclical, seasonal, and irregular elements of prices of paddy, rice, wheat, and bajra. The actual prices of paddy, rice, wheat and bajra are used for the analysis assuming farmers respond to actual prices and not to deflated prices. 12 month moving average is employed assuming seasonal variation has 12 month period.

(2) *Functional Analysis*

Four types of functional analysis were used:

(i) *Analysis of Variance and Correlation Coefficient Matrix*

Analysis of variance is used to study the spatial variation in prices. The zero order correlation coefficient matrix between annual average prices in the markets is calculated to know the relationship between the prices.

(ii) *The Coefficient of Variation*

The coefficient of price variation is worked out for paddy, rice, wheat and bajra for the ten years beginning from 1975-76 to 1984-85. This tool of analysis is used to identify stable markets.

(iii)*Multiple Regression*

A Linear multiple regression model is used showing the relationship between market fee and price index and market arrivals. The multiple regression function is as follows:

$$F_m = a + b_i X_1 + b_2 X_2 + \mu$$

where

F_m is market fee
X_1 is price index

X_2 is market arrivals

a and b are parametres to be estimated.

(iv) *Linear Trends*

Linear trends are fitted, by using the least square method for the market fee and licence fee for the period 1975-76 to 1984-85.

The following hypotheses are tested in the present study:

1. Regulated markets have the features and characteristics of a competitive market.
2. Market regulation creates integration.
3. Market regulation provides stability of markets in terms of prices.
4. The users of the regulated market yards enjoy cost economies.

Conclusion

Thus, as shown in the preceding pages of this chapter, four markets of Kheda district are studied, taking principal crops of the district. The study is made through market structure approach with the help of performance variables and tools of analysis mentioned earlier. The hypotheses mentioned above are tested with the help of primary and secondary data, the sources and the method of collection of which is already discussed. Before examining the working of markets on the lines mentioned in this chapter, the district profile of Kheda district is given in the next chapter so as to get familiarity with Kheda district.

NOTES

1. Lalitha Agricons Private Limited, **Structural Analysis of Operations of Agricultural Produce Market Committee, Unjha,** Gandhinagar: Lalitha Agricons Private Limited, December, 1986.
2. Wayne Purcell, **Agricultural Marketing: System, Coordination, Cash and Future Prices,** Western Virginia: Reston, 1979, p.64.
3. Joe, S. Bain, **Industrial Organisation** New York: John Willey, 1967, p.7.
4. Barbara Harriss, **Paddy and Rice Marketing in Northern Tamil Nadu,** Madras: Sangam Publishers, 1979, pp.85-120.
5. Alice Dewey, **Peasant Marketing in Jawa,** New York: The Free Press, 1969.
6. Willard and Garolan, Bob, Daniel, I, Pedberg, **Economics of Food Retailing,** Ithaca: Cornel University Food Distribution Programme: 1969; O.V. Wells, "Market Structure for Agricultural Development", Proceedings of the Eleventh International Conference of Agricultural Economics 1961 reviewed by Digest of Agricultural Economics and Marketing (No.21, July 1963), p.17; Barbara Harriss, **Paddy and Rice Marketing in Northern Tamil Nadu, op.cit.**
7. G. (Jr.) Bressler Raymond and Richard, A. King **Market, Price and International Trade,** New York: John Wiley & Sons, Inc., 1970, p.73.
8. Z.Y. Jasdanwala,**Marketing Efficiency in Indian Agriculture,** *op. cit.*
9. (i) Uma Lele, **Agricultural Policy in Developing Countries,** ed. Nurul Islam,

London: Macmillan, p.414.

(ii) B.D. Joshi, **Working of Regulated Markets in India,** Vol. 2, Market Series No.175 Department of Agriculture, Government of India, Nagpur.

(iii) M.L. Dantwala, **Report of the Expert Committee 1951** (This Committee was appointed by the Government of Bombay to enquire into Regulated Markets in Bombay in 1950) cited by K, Kulkarni, **Agricultural Marketing in India,** Vol. 2 Bombay: The Cooperators' Book Depot, 1951, p.704.

(iv) C.H. Hanumantha Rao and K. Subbarao, **Marketing of Rice in India : An Analysis of the Impact of Producers' Price on Small Farmers Teaching and Research Forum** No.18 New York: The Agricultural Development Council, May 1970.

10. James V. Rhodes, **The Agricultural Marketing System** Columbia; GRID Publishing, Inc. 1978.
11. P.K. Dave, "Risk and Uncertainty in Indian Agriculture", (Unpublished Ph.D. Thesis), Gujarat University, 1980. p 83.
12. Uma Lele, *op.cit.*
13. James V. Rhodes, *op.cit.*
14. M.L. Dantwala, "Problems before the New Marketing Agencies", **Indian Journal of Agricultural Economics,** 12, April-June, 1957.
15. Barbara Harriss, "How to study Agricultural Marketing and How not to Study it." Madras Institute of Development Studies, Working Paper No.7, 1980. p.41.
16. R.W. (Jr.) Cumings, **Pricing Efficiency in the Indian Wheat Market,** New Delhi, Impex, 1967.
17. Uma Lele, **Foodgrain Marketing in India - Private Performance and Public Policy,** Ithaca : Cornel University Press, 1971, p.89.
18. M.L. Dantwala, *op.cit.,* p.184.
19. (a) V.S. Vyas and M.H. Maharaja, *op.cit.,* pp.52-78. (b) S.S. Acharya, *op.cit.,* p.360.
20. **National Commission on Agriculture, Abridged Report,** New Delhi, Ministry of Agriculture and Irrigation, Department of Agriculture, 1976, p.578.
21. C.T. Wong, *op.cit.*

IV

DISTRICT PROFILE

Kheda district is situated in central Gujarat and it is the fifth smallest district. The total area of the district is 7,194 sq.kms. which is 3.67 per cent of the State's total area. Kheda district lies between 22° - 7′ and 23° - 18′ north latitudes and it is bounded on the north by Sabarkantha district, on the south by river Mahi and Gulf of Cambay, on the east by the district of Panchmahal, south east by Vadodara and west by Ahmedabad district. The Sabarmati and Mahi are the main rivers besides Shedhi, Vatrak, Meshow, Khari, Varansi and Mohar rivers.

There are ten talukas in the district and they are Cambay (Khambhat), Anand, Nadiad, Thasra, Borsad, Matar, Mehmedabad, Petlad, Balasinor, and Kapadvanj.

The district is famous for its Agriculture College and Dairy Science College at Anand. Amul Dairy is a renowned monument. There are plenty of cottage and small scale industries. There are textile mills in Khambhat, Nadiad and Petlad. The cooperative sugar factory and cement works in Pelej and Sevalia are the large scale industries.

In 1901, Kheda had population of 1,037,249. In 1971, the population rose to 2,451,387 (rural 1,962,357 and urban 489,030), recording 157.69 per cent rise during last seventy years. According to the 1981 census, Kheda's population is 3,015,027 out of which the rural population is 2,408,672 and urban population is 606,635. The density of population comes to 419 per sq.km. as against 174 per sq.km. for the State of Gujarat as a whole. The urban density is higher than rural density. Anand taluka has the highest population and highest density (590) in the whole district, because it is an important commercial, agricultural and educational centre. Khambhat has the lowest density (185) because it is not all that developed.

Gujarati is the mother tongue of nearly 95 per cent of the population of Kheda district and Hindus predominate. Kheda's Patel community is spread far and wide. Kheda has given two prominent leaders and both are Patels - they are, Shri Vithalbhai Patel and Shri

Sardar Vallabhbhai Patel.

Regarding pattern of land utilisation in Kheda district 689,139 hectares of land is actually utilised under various uses such as forest, barren and uncultivable land, non-agricultural use, culturable waste, permanent pasture and other grazing land, current fallow land, net area sown etc.

The total cropped area in the district is 6,23,942 hectares out of which 5,23,636 hectares is net area sown and the rest 1,00,306 hectares is area sown more than once.

In Kheda, the main food crops are Paddy, Bajra, Wheat, Gram, Tur whereas non-food crops are Groundnut and Tobacco. The non-food crops are grown for better economic returns. In the year 1982-83, 21.85 per cent of the total cropped area is under Rice, 17.36 per cent is under Kharif Bajra, 5.07 per cent is under summer Bajra, 8.68 per cent is under Wheat, 3.78 per cent is under summer Groundnut and 1.43 per cent is under Kharif Groundnut. The rest 41.83 per cent of land is under other crops.

The land available for cultivation is limited and so the scope of increasing the net cultivated area is limited. Thus intensive cultivation has become important. The intensive cultivation of using agricultural inputs in ideal doses and following scientific methods, introducing high yielding varieties has increased the yield per hectare considerably.

The climate of the district is dry and hot, except in the south-west monsoon season. The period from June to September is the south-west monsoon season. October and November is the post-monsoon season. In 1975-76, during June to September, Kheda experienced 297.07 mm of rainfall and in 1984-85, during the same season, the district had 134.5 mm of rainfall. October to November is the retreating monsoon season in the year 1977-78,in the same month it rained only 22.5 mm.

From January till May, there is hardly any rainfall. The real monsoon begins only in June and after. On an average there are 37 rainy days in a year in Kheda district. The rainy days mean days with rainfall of 2.5 mm - 10 cents or more.

Kheda experiences cyclonic storms from Arabian Sea, gusty winds and widespread rains. Thunder storms occur in the latter half of summer and during monsoon period.

Monsoon is fairly long in the district of Kheda but on the whole it is not adequate and not evenly distributed. Mahi and Sabarmati are the two main rivers of the district and of the two rivers, Mahi is more useful for irrigation.

In Kheda, there are no private canals. However, the main source of irrigation is wells and tubewells. The length of Government canals is 376 kms in the district and among talukas, Petlad has the longest Government canal which is 110 kms. The district has 558 Government tubewells, 490 private tubewells, 232 tanks and 813 government wells and 25630 Kaccha wells.

The total irrigated area in the district is 3,07,244 hectares out of which 2,29,269 hectares is the net irrigated area and 77,975 hectares is the area irrigated more than once. As there are no private canals in Kheda, there is no irrigated area through private canals. However, 1,74,471 hectare area is irrigated by wells, 1,28,348 hectares by Government canals, 4,243 hectares by tanks and 682 hectares by sources other than mentioned above. Anand taluka has the largest area under irrigation followed by Nadiad and Matar.

The principal foodcrops irrigated in the district are Paddy, Wheat and Bajra. These crops are grown in all the talukas of Kheda district. Irrigated tobacco is grown mainly in Anand, Petlad and Borsad taluka and irrigated cotton V-797 is mainly grown in Khambhat taluka.

It is found that in Kheda district 1,02,868 hectares irrigated area is under Rice, 38,834 hectares irrigated area is under Wheat, 31,651 hectares irrigated area is under Bajra. Moreover, 23,601 hectares irrigated area is under tobacco cultivation in the district. However, 2,08,777 hectares is under foodcrops and 98,467 hectares is under nonfood crops.

During 1971 to 1980, minor irrigation schemes viz. Jetholi, Wagas, and Ranipodra were completed and they provide a total gross command area of 3,226 hectares. Three minor irrigation schemes were under construction in 1981 viz. Khata, Bhagvangina Muvada and Debhari. Under these schemes, 687 hectares will be irrigated. Moreover, Mahi Stage-I and Kadana are the medium irrigated schemes and in 1980-81, their achievement was 116.67 hundred hectares.

In 1984-85, the area under paddy was 1,31,900 hectares and it produced 3,912 hundred M.T. and the value of which is 7,307.22 lakh rupees. Bajra was grown in the area of 1,229 hundred hectares. The output was 1,748 hundred M.T. the value of which was 2,636.5 lakh rupees. Wheat was under 503 hundred hectares of area, the output was 849 hundred M.T. and the value of which was 1,715.66 lakh rupees. The area under tobacco was 316 hundred hectares and it produced 462 hundred M.T. output, worth rupees 1,917.30 lakhs.

Cooperative societies have been playing an important role in the rural economy of Kheda district. There is a super structure of cooperative societies with an apex bank at the state level, the District Central Cooperative Bank at the district level and the cooperative societies at the village level. The members of the agricultural credit cooperative societies are provided short and medium term loans for seeds, fertilisers, agricultural implements, etc. The head office of the cooperative societies have increased from 2064 in 1971-72 to 2624 in 1979-80. The scheduled banks have also increased their credit activities during the period 1971-72 to 1979-80.

In the year 1978-79, there were 32 fishing cooperative societies with 605 members.

There were 209 small scale industrial units at the end of 1980. There were 5 Industrial Estates managed by Gujarat Industrial Development Corporation in Kheda in the same year.

There were 135 Joint-Stock Companies as on 1-3-1979, 218 Commercial Banks at 133 places in the district as on 1-3-1979. Moreover, there were 116 Cooperative Banks at 77 places on that date.

The Gujarat Electricity Board provides power in the district and in 1981, 853 villages were electrified.

In 1980-81, the total road mileage was 3084 kms and the length of railway line was 319 kms. All the talukas except Balasinor and Matar are connected by railways. Kheda district has 64 railway stations. There is only one single port namely Khambhat. All the towns in the district of Kheda have the facilities of post, telegraph and telephone facilities. It is amazing to note that, out of 965 inhabited villages, 900 are well connected with State Road Transport buses.

The function of primary education is carried on by the District Panchayats in rural areas and this responsibility is shouldered by the municipalities in urban area.

The district has 30 hospitals, 151 dispensaries, 245 family planning centres, 18 primary health centres, 222 primary health sub-centres, 165 maternity and child welfare centres, and 1,169 community health workers.

There are 15 towns which have hospital facility, 13 towns have dispensary, 10 towns have family planning centres, 6 towns have nursing homes, Pali has a health centre and Petlad and Kapadvanj have T.B. clinics also.

Thus, the district of Kheda is on the whole advanced and prosperous.

Having got the familiarity with Kheda district, the working of selected regulated markets is examined in the next chapter.

V

WORKING OF REGULATED MARKETS: An Analysis

Introduction

In this chapter, an attempt has been made to evaluate the working of the four selected markets of Kheda district, viz. Borsad, Khambhat, Kapadvanj and Mehmedabad. The working of the markets has been examined in terms of the competitive nature of the market, spatial and seasonal integration, and marketing cost inside and outside the regulated market-yard.

According to Jasdanwalla[1] the marketing organisation is said to be working efficiently, if the territorial differences in prices keep within the reasonable limits of the transportation costs. If the differences in prices over time do not exceed storage costs substantially, it means that the organisation is operating efficiently. Marketing structure could be assessed as efficient if it offers agricultural products at prices which reflect realistically the supply and demand situation. In some of the other studies, a common approach to the study of marketing efficiency is that of marketing margins and costs also[2].

Micro-economic theory considers perfect market as the most efficient and therefore an ideal one. Text-books on microeconomic theory lay down the following conditions for the market to be a competitive one. 1) Many buyers, 2) Many sellers, 3) Homogeneous product, 4) Free entry and free exit, 5) Perfect knowledge, 6) Perfect mobility.

Regulated markets come into existence through legal procedure with definite economic and social objectives. Therefore, some additional conditions (given below) become necessary to judge the performance of a regulated market.

(i) Spatial integration of markets

(ii) Controlled oscillations of market prices around normal prices i.e. market stability.
(iii) Absence of excessive profits to the traders.
(iv) Absence of corruption and crime. (v) Amenities to market participants
(vi) Helpfulness of the market staff and functionaries
(vii) Disposal quickness
(viii)Immediate payment
(ix) Pilferage and waste
(x) Market consciousness
(xi) Market information

The Chapter has been divided into four sections. the first section discusses some basic conditions of a competitive market viz. many buyers, many sellers, freedom of entry and exit and impersonal nature of market, (i.e. either the buyer or the seller is ineffective in influencing the market price) and mobility. In the second section, it has been examined whether the markets are integrated or not. The third section examines the problem of market stability where price fluctuations in the short run (i.e. during the seasons) and long-run (i.e. cyclical fluctuations around the trend) have been examined. The fourth section examines (i) cost of buying and selling incurred by the traders and farmers respectively; (ii) producer's price and sale inside and outside regulated market yard. In the last section, marketing efficiency is judged with the help of market arrivals and all the non-economic variables viz., disposal quickness, immediate payment, amenities to market participants, helpfulness of the market staff and functionaries to the transactors, corruption and crime, pilferages and wastes, market consciousness and market information have been discussed.

Section I

CHARACTERISTICS OF A COMPETITIVE MARKET

In this section, the following hypothesis is tested:

"Regulated markets have the features and characteristics of a competitive market."

This hypothesis is tested with the help of number of buyers and sellers participating in the market and also with the freedom of entry and exit on the part of farmers and traders. To prove that there are many buyers and many sellers in the market, the number of lots of notified crops and the number of licenced holders are used following the work Dr. P.N. Kaliyaperumal[3]. This information for the year 1984-85 has been presented in Tables 5.1 and 5.2.

It can be seen from the tables that the notified crops and their number vary from market to market. It has been assumed here that each lot has been brought to the market by a separate individual farmer. In reality, the same farmer might have been visiting the market during the year more than once and therefore, the number of lots cannot be identified with the number of farmers. To that extent, the assumption gives an unrealistic picture. But in practice, it is very much difficult to find out the farmers who have brought their lot to the market more than once and the actual number of lots brought by them to the market. Therefore, the assumption becomes necessary to make the analysis feasible.

TABLE 5.1: Number of Lots in Selected Crops sold in Selected Regulated Market Yards During 1984-85

Market Yard	Paddy	Bajra	Wheat
Borsad	1352	1087	690
Khambhat	9026	4744	4295
Kapadvanj	—	658	380
Mehmedabad	548	189	529
Total	10926	6678	5894

In case of the market for paddy, as shown in Table 5.1, the number of lots in Borsad, Khambhat and Mehmedabad are 1352, 9026 and 548 respectively. The total number of lots for paddy market at four selected locations were 10926. The number is sufficiently big to make the paddy market competitive from sellers' side.

So far as bajra crop is concerned, the lots in the selected markets of Borsad during 1984-85 were 1087, 4744, 658 and 189 respectively (Table 5.1). The total lots (6678) are sufficient to make the market competitive.

In the case of wheat 690, 4295, 380 and 529 lots came for sale in the markets of Borsad, Khambhat, Kapadvanj and Mehmedabad respectively. The total number of lots (5894) for wheat during 1984-85 in selected locations mentioned above are good in number to make the market competitive.

Buyers' side of the market has been analysed with the help of the number of licenced traders (who are the buyers in the market) in the district in 1984-85.

TABLE 5.2: Number of Licence Holders in Four Selected Market Yards During 1984-85

Market Yard	Number of Traders holding licence	Number of Institutions holding licence
Borsad	1856	40
Khambhat	3415	41
Kapadvanj	2363	31
Mehmedabad	1029	14
Total	8663	126

Table 5.2 shows that the number of buyers are sufficiently large in the four selected markets.

Besides individual traders, there are institutions holding licence also. This shows that the number of buyers are big enough to make the market competitive from the buyers' side.

The competitive market has a characteristics of being an impersonal one. In an impersonal market, one particular buyer or a seller is not in a position to affect the market price. From the point of view of sellers' i.e. farmers, their number is also large and the quantity sold by a particular farmer is such a small quantity of total arrivals that he will not be in a position to affect the market price. Further, farmers come from scattered villages of the area and therefore there is much less possibility of knowing one another and making collective moves in the selling of their produce. Thus oligopolistic tendencies are less likely to arise on selling side. But in case of traders who buy the produce are in a position to have tacit understanding in the purchase of the produce and it is quite possible that one or few traders might purchase either the whole or a very large portion of the produce. So, oligopsony is likely to arise. In order to know whether, oligopsony arises in agricultural marketing or not, the concept of "buyer concentration ratio" has been used.

TABLE 5.3: Buyer Concentration for Paddy on Four Selected Dates* in Four Selected Market Yards During 1984-85

Market Yard	Code (1)**	(2)**	(3)**	(4)**	(5)**	(6)**
Borsad	54	618	10	3	300	618
Khambhat	304	1717	26	1	164	1717
Kapadvanj	-	-	-	-	-	-
Mehmedabad	23	165	10	3	85	165

* 8-11-84, 15-11-84, 22-11-84, 29-11-84

**Code:
(1) Total number of lots on four selected dates.
(2) Total number of bags on four selected dates.
(3) Total number of traders who purchased on four selected dates.
(4) Total number of traders who purchased 50% or more on four selected dates.
(5) Total quantity purchased by those traders (in quintal) on four selected dates.
(6) Total quantity purchased (in quintal) on four selected dates.

Source: Data were collected from the concerned regulated markets.

TABLE 5.4: Buyer Concentration for Wheat on Four Selected Dates* in Four Selected Market Yards During 1984-85

Market Yard	Code (1)**	(2)**	(3)**	(4)**	(5)**	(6)**
Borsad	64	490	10	2	52	490
Khambhat	162	1224	44	1	9	1224
Kapadvanj	21	1.75	10	3	0.87	1.75
Mehmedabad	33	455	14	3	189	455

* 8-3-85, 15-3-85, 22-3-85, 29-3-85

**Code:
(1) Total number of lots on four selected dates.
(2) Total number of bags on four selected dates.
(3) Total number of traders who purchased on four selected dates.
(4) Total number of traders who purchased 50% or more on four selected dates.
(5) Total quantity purchased by those traders (in quintal) on four selected dates.
(6) Total quantity purchased (in quintal) on four selected dates.

Source: Data were collected from the concerned regulated markets.

Table 5.3 to 5.5 give cropwise information pertaining to number of lots, number of bags, number of traders who purchase, number of traders who purchased 50 per cent or more of the quantity.

TABLE 5.5: Buyer Concentration for Bajra on Four Selected Dates* in Four Selected Market Yards During 1984-85

Market Yard	Code (1)**	(2)**	(3)**	(4)**	(5)**	(6)**
Borsad	16	64	4	4	64	64
Khambhat	74	616	24	1	29	616
Kapadvanj	20	1.40	5	4	1.36	1.40
Mehmedabad	3	28	3	3	28	28

* 8-5-85, 15-5-85, 22-5-85, 29-5-85.

**Code: (1) Total number of lots on four selected dates.
(2) Total number of bags on four selected dates.
(3) Total number of traders who purchased on four selected dates.
(4) Total number of traders who purchased 50% or more on four selected dates.
(5) Total quantity purchased by those traders (in quintal) on four selected dates.
(6) Total quantity purchased (in quintal) on four selected dates.

Source: Data were collected from the concerned regulated markets.

This information is given on the selected dates of a selected months of the year 1984-85. The selected months are not uniform for all crops. Paddy is kharif crop and so, November is the month selected for paddy, because major portion of the annual arrival of paddy comes to the market in the month of November. Wheat is a winter crop and so major portion of annual arrival of wheat comes to the market in the month of March. Therefore, March is the selected month for wheat crop. In case of bajra, it is both a kharif and a summer crop. However, in the selected markets, monthwise arrival of bajra shows that a large portion of annual arrival of bajra comes to the market in the month of May except Khambhat market yard. To maintain uniformity of month for all the markets including Khambhat, the month of May has been taken for studying bajra.

So far as paddy is concerned, the total number of lots in Borsad, Khambhat and Mehmedabad market yards were 54,304 and 23 respectively. The total number of bags purchased in the markets were 618, 1717 and 165. The number of purchasers (traders) who purchased the quantity sold was very high and so the markets were competitive.

In case of wheat, the total number of lots in Borsad, Khambhat, Kapadvanj and Mehmedabad market yards are 64, 162, 21 and 33 respectively. The total number of bags purchased in the market yards were 490, 1224, 1.75 and 455. The number of purchasers (traders) who purchased the total quantity sold was very high and so the markets were competitive.

So far as bajra is concerned, the total number of lots in Borsad, Khambhat, Kapadvanj and Mehmedabad market yards were 16, 74, 20 and 3. The total number of bags purchased in the market were 64, 616, 1.40 and 28. The number of purchasers (traders) who purchased the quantity sold was very high and so the markets can be said to be competitive.

Here, buyer concentration ratio is calculated with a view to find out whether the market is competitive or not. In other words, it helps in finding out the presence or absence of monopolistic or oligopolistic tendencies in the market. It can be calculated as mentioned below:

$$\text{Buyer concentration ratio} = \frac{\text{Number of traders who purchased 50\% or more}}{\text{Number of total traders}} \times 100$$

In Borsad market yard, 618 quintals of paddy were purchased by 10 traders[4]. The number of traders who purchased 50% or more on particular dates was 3 and their total purchase on the four dates of the month was 300 quintals of paddy. Thus, 30% of traders purchased 48.54% of the purchase. In Khambhat market, 1717 quintals of paddy were purchased by 26 traders. The number of traders who purchased more than 50% on particular dates was 1 and their total purchase on the four selected dates of the month was 164 quintals of paddy. Thus, 3.85% of the traders purchased 9.55% of the total purchase. In Mehmedabad market yard, 165 quintals of paddy were purchased by 10 traders. The number of traders who purchased more than 50% was 3 and their total purchase on the four selected dates of month was 85 quintals. Thus 30% of the traders purchased 51.5% of the total purchase and the buyer concentration is not very much alarming.

In case of wheat in Borsad market, 490 quintals were purchased by 10 traders. The number of traders who purchased 50% or more on particular dates was 2 and their total purchase on the four dates of the month was 52 quintals of wheat. Thus 20% of the traders purchased 10.61% of the purchase. In Khambhat market yard, 1224 quintals of wheat were purchased by 44 traders. The number of traders who purchased 59% or more on particular dates was 1 and his total purchase on the four dates of the month was 9 quintals of wheat. Thus in Khambhat market yard 2.27% of the buyers purchased 0.73% of the purchase. In Kapadvanj market yard, 10 traders purchased 1.75 quintals of wheat. There were 3 traders who purchased 50% or more on particular dates and their total purchase on the four dates of the month was 0.87 quintals. Thus, 30% of the traders purchased 49.71% of the total purchase. In Mehmedabad market, 14 traders purchased 455 quintals of wheat. There were 3 purchasers who purchased 189 quintals. Hence, 21.43% of purchasers purchased 41.54% of wheat. The buyer concentration is not very high in case of wheat.

In Borsad market yard, 64 quintals of bajra were purchased by 4 traders. The number of traders who purchased 50% or more on particular dates was also 4 and their total purchase on the four dates of the month was 64 quintals. Hence, 100% of traders purchased 100% of the quantity. In Khambhat market yard, 616 quintals of bajra were purchased by 24 traders, and one trader purchased more than 50%, on particular date and his total purchase was 29 quintals. Thus, 4.17% of the traders purchased 4.17 quintals of bajra. In Kapadvanj market yard, 5 traders purchased 1.40 quintals of bajra on particular dates. There were 4 traders who purchased 50% or more and they purchased 1.36 quintals. Thus, in Kapadvanj market yard, 80% of the purchasers purchased 97.14% of the total purchase of bajra. In Mehmedabad, the buyer concentration ratio is 100%. There were 3 traders who purchased the entire quantity of 28 quintals of bajra. Hence, 100% of traders purchased 100% of the quantity. In case of bajra, the buyer concentration ratio is quite alarming, especially in Mehmedabad, Borsad and Kapadvanj market yards.

Bajra crop shows high buyer concentration ratio in Borsad, Kapadvanj and Mehmedabad market yards. If one looks at the data carefully, it is observed that the amount bought and sold on all selected dates is very small. So, it is quite natural that it will be bought by either one or two traders on all the four dates. As a result, the percentage of traders who bought more than 50% of total traders will be very high (100%) but from this one cannot infer that there is absence of competition and either monopolistic or oligopolistic tendencies prevail in the market in general.

The inference of the above discussion is that in general there are neither oligopolistic nor oligopsonistic tendencies in these market yards. In other words, neither the buycr nor the seller can influence the market price. Thus the markets are competitive and impersonal in nature and maintain the exchange neutrality.

This analysis is restricted to the four selected dates of a selected month. The more realistic picture can be obtained by taking extended time span.

Freedom of Entry and Exit

One of the important characteristics of a competitive market is the freedom of entry and exit. In a market the entry and the exit of the firm should not be restricted by institutional regidities[5]. The regulated markets are established mainly for protecting the interests of the farmers because they are exploited by the traders. So, in study of regulated market yards, freedom of entry and exit of the farmer is more relevant. As per the rules, any farmer can bring his produce to the market yard. The only requirement on the part of a farmer is to obtain gate pass. This gate pass requirement is there to bring smooth functioning of the market. He has a freedom of exit also. After the auctioning of his produce, if the final price of his produce is not acceptable to him, he has a freedom to withdraw his produce and can sell it in future at any time. The empirical data of the withdrawal of his produce from the market is not available for all the market yards. Only for Borsad market yard, the cropwise figures for arrival and sale are available. They are presented in Table 5.6. The difference between the arrival and sales shows that the farmers have withdrawn their crops and thus it can be proved that there is a freedom of exit also.

TABLE 5.6: Arrival and Sale of Selected Crops in Borsad Market Yard During 1984-85

(In quintals)

Crop	Arrival	Sale
Paddy	186817	13443
Wheat	48530	2061
Bajra	81087	7728

Source: Annual Reports of the concerned regulated market yard.

The traders also have the freedom of entry and exit. They are free to purchase different quantities at different prices on the basis of the quality of the produce. Further, the holding of a licence on the part of a trader is actually not a barrier to the entry but it is necessary to regulate his behaviour and to facilitate the functioning of the market.

Mobility

Mobility is also one of the essential characteristics of a competitive market. Actually there are no barriers to the mobility of the produce from the market yards to other outside areas and from outside areas to the market yards. Some information is available regarding mobility of the produce and it is presented in Tables 5.7 to 5.10. The information regarding total stocks, total arrivals etc. (as shown in Table 5.7) on the first Friday of the month during the year 1984-85 shows that in Borsad market yard in case of paddy, arrival is both from local villages and from other markets. However, the outgoing quantities show that they are mainly despatched to outstation areas. In Khambhat market yard, paddy reaches the yard both from local and outstation areas but the despatches are mainly to local areas. In Mehmedabad market yard, the arrival and despatch both involve local and outstation areas but surprisingly the amount for local use remains constant (10 quintals) on all the dates.

In case of rice as shown in Table 5.8 the market arrival is mainly from outstation areas. It is despatched both from local and outstation areas but the local use is more. In Khambhat market yard, arrival of rice from other markets is more and is despatched to both local and outstation areas but the outstation despatch is relatively more. In Mehmedabad market yard, arrival of rice is both from local villages and from other markets but arrival is relatively more from villages than from other markets. So far as despatch is concerned, despatches to outstation area is relatively more and further the amount for local use remains constant (10 quintals) on all the dates. In Kapadvanj market yard, arrival of rice is mainly from villages and despatches are also for local use.

Coming to the mobility of wheat as shown in Table 5.9, arrival from outstation is more in Borsad market yard and despatches are relatively more to outstation areas. In Khambhat market yard, the arrival of wheat is mainly from villages. Despatches of wheat are to both the local villages and outstation areas. In Kapadvanj, barring one

TABLE 5.7: Stocks, Arrivals, Outgoing Quantities of Paddy on the First Friday of the Month in Selected Market Yards During 1984-85

(*In Quintals*)

Market Yard	Stock at the beginning of the week	Arrival during the week			Outgoing Quantities during the week out of of the quantities shown in columns 2 and 5			Stock at the end of the week
		From villages	From other markets	Total	For local use	Despatches outstation	Total	
(1)	(2)	(3)	(4)	(3)+(4)=(5)	(6)	(7)	(6)+(7)=(8)	(9)
Borsad	77651	7957	4505	12462	193	13730	13923	76190
Khambhat	367875	10469	56737	67206	58204	1004	59208	375873
Kapadvanj	-	-	-	-	-	-	-	-
Mehmedabad	14970	850	590	1440	120	1360	1480	14930

Source: Data were collected from the concerned Regulated Markct Yard.

TABLE 5.8: Stocks, Arrivals, Outgoing Quantities of Rice on the First Friday of the Month in Selected Market Yards During 1984-85

Market Yard	Stock at the beginning of the week	Arrival during the week			Outgoing Quantities during the week out of of the quantities shown in columns 2 and 5			Stock at the end of the week
		From villages	From other markets	Total	For local use	Despatches outstation	Total	
(1)	(2)	(3)	(4)	(3)+(4)=(5)	(6)	(7)	(6)+(7)=(8)	(9)
Borsad	20694	1713	6610	8323	8328	1994	10322	19665
Khambhat	131534	227	31825	32052	8059	18097	26156	137430
Kapadvanj	2000	115	949	1064	1764	-	1764	1300
Mehmedabad	15630	890	570	1460	120	1260	1380	15718

Source: Data were collected from the concerned Regulated Market Yard.

TABLE 5.9: Stocks, Arrivals, Outgoing Quantities of Wheat on the first Friday of the Month in Selected Market Yards During 1984-85

Market Yard	Stock at the beginning of the week	Arrival during the week			Outgoing Quantities during the week out of of the quantities shown in columns 2 and 5			Stock at the end of the week
		From villages	From other markets	Total	For local use	Despatches outstation	Total	
(1)	(2)	(3)	(4)	(3)+(4)=(5)	(6)	(7)	(6)+(7)=(8)	(9)
Borsad	11586	3132	3383	6515	776	5458	6234	11867
Khambhat	6585	4907	1207	6114	2668	1831	4499	8200
Kapadvanj	-	2822	32	2854	1204	1650	2854	-
Mehmedabad	12610	690	290	980	120	790	910	12680

Source: Data were collected from the concerned Regulated Market Yard.

TABLE 5.10: Stocks, Arrivals, Outgoing Quantities of Bajra on the First Friday of the Month in Selected Market Yards During 1984-85

Market Yard	Stock at the beginning of the week	Arrival during the week			Outgoing Quantities during the week out of of the quantities shown in columns 2 and 5			Stock at the end of the week
		From villages	From other markets	Total	For local use	Despatches outstation	Total	
(1)	(2)	(3)	(4)	(3)+(4)=(5)	(6)	(7)	(6)+(7)=(8)	(9)
Borsad	40955	6942	4302	11244	1555	11405	12960	39239
Khambhat	6275	9170	-	9170	1868	7770	9638	5807
Kapadvanj	-	5008	66	5074	1199	3875	5074	-
Mehmedabad	12130	650	330	980	120	920	1040	12070

Source: Data were collected from the concerned Regulated Market Yard.

exception, the arrival of wheat is from local villages. But the despatches involve both local and outstation areas. Similarly, in Mehmedabad market yard, arrival of wheat from villages is relatively more and despatches are relatively more to outstation areas.

In case of bajra, in Borsad market yard the arrival is both from villages and other markets (Table 5.10). The despatches are more to outstation than for the local use. In Khambhat market yard, the arrival of bajra is from the villages. The despatches of bajra is relatively more to outstation areas. In Kapadvanj market yard also, the arrival is mainly from the villages, and despatches are relatively more to outstation areas. In Mehmedabad market yard, bajra arrival is relatively more from the villages than from the other markets. The despatches are more to outstation areas than for local use.

Thus the mobility of the produce is much higher and that is essential for a market to be competitive.

Thus, as seen in the above discussion, there are many buyers, many sellers, impersonal nature of market (i.e. neither any buyer nor any seller is in a position to influence the price) and sufficiently high degree of mobility of the produce of the selected crops in the selected market yards of Kheda district. Therefore, the hypothesis:

> "Regulated markets have the features and characteristics of a competitive market"

is found valid.

Section II

MARKET INTEGRATION

One of the important indicators of efficient functioning of market is integration of markets.

According to Uma Lele[6], market integration is the interrelationship between price movements in two markets. The degree of correlation can be taken as an indicator of the extent to which the markets are integrated.

If the markets located at different places are not integrated, there will be wide differences among the prices of different markets. As a result, farmers selling in markets having higher prices will benefit and farmers selling in markets having lower prices will lose. Moreover, the farmer might think at the time of sale that the price in the other market might be higher. Theoretically, price difference between any two markets can be there but it should not exceed the transport cost. In general, what is required is that markets should be integrated over space and time.

Cummings[7] is of the opinion that market integration implies "the association of prices" between markets.

Association of prices overtime shows seasonal integration of prices and association of prices between markets shows spatial integration of prices.

Spatial Integration

Markets are integrated over space when the territorial price difference do not exceed the transport costs plus the remuneration for the services of the trader[8]. This is called spatial integration.

Kumar and Singh, Deole and others[9] are of the opinion that the market can be said to spatially integrated if the regional variations in prices are consistent with the transport costs.

Spatial integration of market yards has been studied here through analysis of variance. To study spatial price variations, monthly annual average prices for ten years have been taken for the period 1975-76 to 1984-85 for four market yards viz. Borsad, Khambhat, Kapadvanj and Mehmedabad. The crops selected for the study are paddy, rice, wheat and bajra. The results are presented in Tables 5.11 to 5.14.

It can be seen from tables 5.11 to 5.14 that variations in prices in all the crops are insignificant. Price variations are there between the markets but they being insignificant, it can be inferred that markets are spatially integrated and insignificant price variations might be due to chance only.

TABLE 5.11: Analysis of Variance for Paddy

Source	SS	DF	MS	F
Between	15191.15	3	5063.72	2.6845
Within	67904.67	36	1886.24	
Total	83095.82	39		

TABLE 5.12: Analysis of Variance for Rice

Source	SS	DF	MS	F
Between	24275.80	3	8091.93	1.3462
Within	216393.72	36	6010.94	
Total	240669.52	39		

TABLE 5.13: Analysis of Variance for Wheat

Source	SS	DF	MS	F
Between	1714.24	3	571.41	0.5189
Within	39639.17	36	1101.08	
Total	41353.41	39		

TABLE 5.14: Analysis of Variance for Bajra

Source	SS	DF	MS	F
Between	66.45	3	22.15	0.04457
Within	17886.96	36	496.86	
Total	17953.41	39		

Market integration over space has been verified also through correlation of prices between markets. Tables 5.15 to 5.18 give correlation matrices for prices of selected crops and show the correlation of prices of different markets. Uma Lele[10] is of the opinion that a correlation coefficient of 0.90 or more is a high degree of inter-market price relationship. Thus, the value of the correlation coefficient in all the tables are significant. Similar results have been obtained by Acharya S.S., and Agarwal, N.L.[11] in his study. In other words, prices of selected crops between the markets are significantly correlated and hence it can be inferred that the markets are integrated.

Thus, the hypothesis:

"Market regulation creates integration of markets" is found valid in the study of selected markets of Kheda.

TABLE 5.15: Correlation Matrix for Prices of Paddy

Between Market Yards	Borsad (X_1)	Khambhat (X_2)	Kapadvanj (X_3)	Mehmedabad (X_4)
Borsad (X_1)	1	0.9458	0.9500	0.9335
Khambhat (X_2)	0.9458	1	0.9160	0.9022
Kapadvanj (X_3)	0.9500	0.9160	1	0.9346
Mehmedabad (X_4)	0.9335	0.9022	0.9346	1

TABLE 5.16: Correlation Matrix for Prices for Rice

Between Market Yards	Borsad (X_1)	Khambhat (X_2)	Kapadvanj (X_3)	Mehmedabad (X_4)
Borsad (X_1)	1	0.9802	0.9520	0.9904
Khambhat (X_2)	0.9802	1	0.9631	0.9737
Kapadvanj (X_3)	0.9520	0.9631	1	0.9557
Mehmedabad (X_4)	0.9904	0.9737	0.9557	1

TABLE 5.17: Correlation Matrix for Prices of Wheat

Between Market Yards	Borsad (X_1)	Khambhat (X_2)	Kapadvanj (X_3)	Mehmedabad (X_4)
Borsad (X_1)	1	0.9928	0.9500	0.9967
Khambhat (X_2)	0.9936	1	0.9949	0.9972
Kapadvanj (X_3)	0.9936	0.9949	1	0.9981
Mehmedabad (X_4)	0.9967	0.9972	0.9981	1

TABLE 5.18: Correlation Matrix for Prices of Bajra

Between Market Yards	Borsad (X_1)	Khambhat (X_2)	Kapadvanj (X_3)	Mehmedabad (X_4)
Borsad (X_1)	1	0.9965	0.9944	0.9961
Khambhat (X_2)	0.9965	1	0.9910	0.9964
Kapadvanj (X_3)	0.9944	0.9910	1	0.9960
Mehmedabad (X_4)	0.9961	0.9964	0.9960	1

Section III

MARKET STABILITY

Stability of a market is an another characteristic of good and efficient market. If the market exhibits price stability,it helps in reducing or avoiding speculation. This aspect has been studied through (i) Coefficient of variation, (ii) Trends, (iii) Cyclical variation and (iv) Seasonal variation.

(i) *Coefficient of Variation*

Wide fluctuations in prices make the market unstable and it hinders the efficient functioning of a market. Stability of prices is an essential characteristic of a sound market. Hence, by estimating the coefficient of variation of prices over a period of ten years, the relatively stable markets have been identified.

Table 5.19 gives coefficient of price variation for paddy, rice, wheat and bajra in four market yards viz. Borsad, Khambhat, Kapadvanj and Mehmedabad for the period 1975-76 to 1984-85. Yearwise values of coefficient of price variation do not give clear picture but coefficient of variation for the whole reference period show that these values are lowest in Mehmedabad market yard and highest in Borsad market yard for all the crops. Thus, Mehmedabad market yard is the most stable market and Borsad is the most unstable one. If we rank them in terms of stability, their order can be Mehmedabad, Kapadvanj, Khambhat, and Borsad. In other words, next to Mehmedabad which is the most stable market, comes Kapadvanj. Further, Kapadvanj is relatively more stable as compared to Khambhat. The last and the least stable market yard is Borsad.

(ii) *Trends*

Stability analysis has been furthered by fitting linear trends in the prices of paddy, rice, wheat and bajra in the four selected market yards. The results are presented in Table 5.20. The lowest trend-coefficient for paddy is 1.07 for Borsad and the highest is 1.23 for Mehmedabad yard. In case of rice, the lowest trend-coefficient is 1.30 for Kapadvanj yard and the highest is 2.54 for Mehmedabad yard. Similarly, the lowest in wheat crop is 0.73 for Borsad yard and the highest is .86 for Khambhat yard. For bajra, the lowest trend-coefficient is .47 for Kapadvanj yard and the highest is 0.58 in Mehmedabad yard.

TABLE 5.19: Coefficients of Price Variation for Paddy, Rice, Wheat and Bajra from 1975-76 to 1984-85 for Borsad, Khambhat. Kapadvanj and Mehmedabad

Year	Borsad				Khambhat				Kapadvanj				Mehmedabad			
	Paddy	Rice	Wheat	Eajra	Paddy	Rice	Wheat	Bajra	Paddy	Rice	Wheat	Bajra	Paddy	Rice	Wheat	Bajra
1975-76	9.47	3.94	22.32	8.70	6.24	15.36	19.72	9.56	12.34	8.84	22.42	12.64	8.73	6.86	21.49	8.06
1976-77	13.32	13.72	4.89	11.93	9.81	8.71	6.17	7.72	11.96	7.32	5.54	9.58	12.76	10.12	4.92	12.34
1977-78	5.83	4.03	9.96	12.78	3.43	10.41	4.89	13.96	5.16	4.69	12.98	16.23	5.48	5.35	10.11	12.89
1978-79	6.35	0.60	5.48	5.35	16.02	11.25	4.91	5.55	8.94	7.03	6.16	5.43	11.52	8.38	7.74	6.93
1979-80	4.35	8.00	7.81	8.83	17.76	11.83	8.46	8.51	24.94	8.55	8.64	8.87	3.45	3.71	8.56	5.50
1930-81	13.74	7.81	7.26	10.30	8.95	5.21	9.03	9.42	11.79	10.76	6.87	11.50	9.29	2.67	6.68	9.20
1931-82	12.44	12.73	9.62	6.23	10.76	11.65	3.89	4.04	14.49	13.89	8.44	6.03	10.82	8.36	8.01	6.72
1932-83	5.11	9.94	8.88	6.55	5.74	5.10	10.49	9.36	10.51	8.19	6.70	9.88	9.02	8.19	6.26	8.69
1933-84	12.75	9.44	12.42	8.67	4.58	8.35	8.28	11.81	7.99	7.83	12.88	9.89	6.29	5.40	11.62	6.81
1934-85	12.28	8.58	11.11	17.94	10.34	6.68	9.76	17.09	5.77	4.33	11.41	20.65	12.83	5.57	9.54	14.92
1975-85	28.75	25.62	23.00	19.81	26.77	25.53	22.76	19.47	26.72	19.84	22.00	19.23	25.94	27.61	21.90	19.59

TABLE 5.20: Linear Price Trend of Paddy, Rice, Wheat and Bajra

Sr. No.	Name of Market	Trend Equation			
		Paddy	Rice	Wheat	Bajra
1.	Borsad	83.90 + 1.0766X	118.10 + 1.8398X	118.06 + 0.7368X	102.02 + 0.5218X
2.	Khambhat	116.70 + 1.0957X	181.44 + 1.8577X	126.46 + 0.8688X	102.65 + 0.5489X
3.	Kapadvanj	104.05 + 0.8809X	226.30 + 1.3057X	114.39 + 0.8211X	104.37 + 0.4734X
4.	Mehmedabad	115.77 + 1.2360X	198.82 + 2.5473X	114.91 + 0.8377X	100.42 + 0.5854X

Origin, X = 0 at October

X units : 1 month.

(iii) *Cyclical Variations*

Absence of cyclical variations in prices is essential for a stable market. If prices vary cyclically, there will be booms and depressions and as a result, there will be all the evil consequence of a business cycle. Agricultural prices experience cyclical variations in prices because of good and bad years. In good years, there is bumper crop and so agricultural prices come down. Similarly, in bad years, the supply is less and, therefore, prices go up. The average length of the cycle varies.

The deviation of the actual price from the trend value in each year has been calculated in percentage terms. This concept is alternatively known as "cyclical relative" and the same has been used in subsequent discussion. With its help the highest and the lowest values in each crop and each market are discussed below so as to understand the intensity and simultaneity of a cycle.

The computed percent of trend (cyclical relative) for the crops selected in the present study are given in Tables 5.21 to 5.24.

In case of paddy crop, Borsad and Mehmedabad market yards experienced the highest cyclical relatives (116.69 and 115.99) in the same year i.e. 1975-76. The lowest cyclical relative (83.95 and 87.90) for the corresponding crop in the corresponding markets yards were experienced in the year 1978-79, whereas Khambhat and Kapadvanj market yards experienced highest cyclical relatives (121.51 and 122.73) in the same year i.e. 1982-83 and the lowest cyclical relatives (87.36 and 78.94) in the same year i.e. 1977-78.

So far as rice is concerned, Borsad and Mehmedabad market yards experienced the highest cyclical relatives (119.05 and 121.90) in the same year i.e. 1976-77 and the lowest cyclical relatives (86.44 and 86.89) in the same year i.e. 1978-79. In Khambhat and Kapadvanj market yards, the highest and the lowest cyclical relatives for rice are experienced in different years.

Regarding wheat, it is observed that the highest and the lowest cyclical relatives are scattered in different years in different market yards.

It has been observed in bajra crop that the highest cyclical relatives in all the four market yards, viz. Borsad, Khambhat, Kapadvanj and Mehmedabad are 111.24, 112.59, 112.46, and 111.89 respectively in the same year i.e. 1980-81. The lowest cyclical relatives for bajra are 80.84, 79.21, 82.82 and 79.81 for the respective market yards in 1978-79.

TABLE 5.21: Cyclical Relatives of Paddy Prices in Four Selected Market Yards for the Period 1975-76 TO 1984-85

Year	Borsad	Khambhat	Kapadvanj	Mehmedabad
1975-76	116.69	103.22	118.15	115.99
1976-77	113.77	104.42	120.05	114.36
1977-78	92.87	87.36	78.94	94.13
1978-79	83.95	97.23	80.85	87.90
1979-80	93.26	98.61	102.08	91.77
1980-81	91.52	98.58	90.20	89.58
1981-82	95.95	103.08	92.01	92.85
1982-83	115.68	121.51	122.73	109.82
1983-84	106.44	97.61	103.77	110.15
1984-85	94.41	88.18	94.25	97.51

TABLE 5.22: Cyclical Relatives of Rice Prices in Four Selected Market Yards for the period 1975-76 to 1984-85

Year	Borsad	Khambhat	Kapadvanj	Mehmedabad
1975-76	119.05	111.53	119.73	121.80
1976-77	113.11	111.02	129.58	109.64
1977-78	94.49	95.02	139.43	96.18
1978-79	86.44	95.53	149.29	86.89
1979-80	91.69	85.49	159.14	95.37
1980-81	88.81	88.39	168.99	87.73
1981-82	90.03	97.40	178.85	87.56
1982-83	112.89	112.40	188.70	109.32
1983-84	103.85	105.22	198.55	109.12
1984-85	103.57	99.90	208.41	101.77

TABLE 5.23: Cyclical Relatives of Wheat Prices in Four Selected Market Yards for the Period 1975-76 to 1984-85

Year	Borsad	Khambhat	Kapadvanj	Mehmedabad
1975-76	109.55	109.21	107.45	111.28
1976-77	96.34	97.78	113.13	96.55
1977-78	99.16	101.27	118.81	98.80
1978-79	87.29	86.90	124.49	87.23
1979-80	92.69	91.99	130.17	91.75
1980-81	100.88	101.04	135.85	100.48
1981-82	107.40	105.58	141.54	106.68
1982-83	114.35	112.86	147.22	112.94
1983-84	100.47	98.12	152.90	100.49
1984-85	91.56	95.25	158.58	93.61

TABLE 5.24: Cyclical Relatives of Bajra Prices in Four Selected Market Yards for the Period 1975-76 to 1984-85

Year	Borsad	Khambhat	Kapadvanj	Mehmedabad
1975-76	98.16	99.25	98.20	98.76
1976-77	101.79	103.42	102.25	104.29
1977-78	107.01	104.61	103.86	103.90
1978-79	80.84	79.21	82.82	79.80
1979-80	102.06	100.27	103.13	101.65
1980-81	111.24	112.59	112.46	111.89
1981-82	104.91	105.93	104.92	106.31
1982-83	102.77	104.49	101.07	103.11
1983-84	93.09	93.08	91.56	91.22
1984-85	98.14	97.18	99.71	99.15

(iv) *Seasonal Variations*

The basic difference between industry and agriculture is that industrial production can go on all throughout the year, whereas in case of agriculture, its activity is normally restricted to a particular season of the year. Thus the seasonality of agriculture results into fluctuations in the supply of agricultural produce during the year. Immediately, after the harvest, the supply increases in the market and therefore, prices go down during that period. In the remaining part of the year, the supply decreases and therefore, prices go up. However, this is not hundred per cent true because time has changed and now the farmer has the capacity and the facility to store and preserve the produce till the market prices go up to his expectations. Credit facilities and storage facilities either in his own house or in any warehouse has changed the situation. But still there are some farmers especially the marginal and small ones who dispose of their produce either because they need cash or because they are forced to sell to the money lender or because they do not have the storage facility. All these result into uneven supply in the market over the months of a year and consequently, there are price fluctuations over seasons in a given year. The seasonal indices of prices of the selected crops in the selected markets have been worked for the period 1975-76 to 1984-85 and are presented in tables 5.25 to 5.28. Their graphical presentation is given in figures 1 to 16 (Appendix II).

TABLE 5.25: Seasonal Indices of Prices of Paddy in Selected Market Yards During 1975-76 to 1984-85

Months	Borsad	Khambhat	Kapadvanj	Mehmedabad
October	93.29	101.91	92.73	99.24
November	91.96	96.35	90.46	91.76
December	97.42	99.34	92.98	96.97
January	98.68	100.63	96.74	101.07
February	97.42	99.24	98.31	97.10
March	97.24	95.22	100.67	97.19
April	98.91	96.13	108.67	97.04
May	101.33	98.90	99.37	97.43
June	100.38	101.88	102.61	103.00
July	111.12	102.63	105.13	105.93
August	108.91	103.31	104.07	106.85
September	103.32	104.45	108.25	106.41

It can be seen from Table 5.25 that the prices of paddy are lowest in the month of November in all the four market yards. This is because November is the post-harvest month and the market arrivals are more. The price indices of paddy in the month of November in Borsad, Khambhat, Kapadvanj and Mehmedabad market yards are 91.96, 96.35, 90.46 and 91.76 respectively. So far as the month of highest price is concerned, there is no single month for all the market yards. However, in general, it can be said that paddy prices are high in the off-season i.e. during June to September. The direction and duration of price change is not uniform in the four selected market yards. It can be seen from figures 1, 2, 3 and 4 also (Appendix II).

TABLE 5.26: Seasonal Indices of Prices of Rice in Selected Market Yards During 1975-76 to 1984-85

Months	Borsad	Khambhat	Kapadvanj	Mehmedabad
October	104.31	105.44	101.54	98.95
November	92.76	109.96	90.62	97.00
December	98.02	109.31	99.19	99.80
January	97.66	104.13	100.81	100.35
February	96.92	97.92	101.61	99.63
March	97.93	93.93	99.43	98.20
April	98.39	92.12	98.37	97.95
May	100.67	90.98	99.25	98.48
June	102.38	94.94	104.42	100.84
July	104.36	96.75	102.40	102.51
August	105.12	100.31	100.79	104.10
September	101.48	104.19	101.59	102.19

TABLE 5.27: Seasonal Indices of Prices of Wheat in Selected Market Yards During 1975-76 to 1984-85

Months	Borsad	Khambhat	Kapadvanj	Mehmedabad
October	99.17	95.74	99.59	100.55
November	102.82	100.89	107.66	107.08
December	109.59	104.60	109.04	110.45
January	110.29	103.67	107.30	109.36
February	106.15	102.52	105.83	105.32
March	94.99	104.30	94.28	94.24
April	91.79	104.12	92.61	90.80
May	91.90	101.67	89.65	90.01
June	93.16	96.27	93.50	92.05
July	99.26	94.39	97.57	97.19
August	101.19	96.18	102.46	102.19
September	99.68	95.65	100.51	100.74

TABLE 5.28: Seasonal Indices of Prices of Bajra in Selected Market Yards During 1975-76 to 1984-85

Months	Borsad	Khambhat	Kapadvanj	Mehmedabad
October	89.38	89.12	90.78	90.07
November	94.81	97.89	98.64	95.60
December	101.29	99.44	99.64	100.26
January	102.71	102.82	104.63	101.78
February	102.82	104.63	105.99	103.42
March	105.58	106.70	102.58	105.58
April	107.66	104.89	98.75	104.30
May	101.04	100.40	100.67	100.31
June	98.15	98.99	99.78	97.64
July	102.02	101.02	102.44	102.93
August	101.59	99.90	101.68	103.41
September	92.96	94.19	94.43	94.72

In case of rice as found in Table 5.26, the lowest price is found in general in the month of November. The indices of the price of rice in Borsad, Kapadvanj and Mehmedabad market yards are 92.76, 90.62, and 97.00 respectively. Khambhat market yard is an exception to it. Khambhat market yard experienced the lowest price of rice in the month of May and the price index in that month was 90.98. There is no uniformity with regard to the month of highest price. In Borsad and Mehmedabad market yards, August is the month of highest price and the price indices are 105.12 and 104.10 respectively. In Khambhat market yard, the highest price is found in the month of November (109.96) and in Kapadvanj market yard it is found in June (104.42). Thus, there is no uniformity of month as far as the highest price is concerned but it prevails in the off-season i.e. during May to September. This can be seen from the figures 5, 6, 7 and 8 also (Appendix II).

Wheat is a winter crop. The crop comes to the market in the latter part of March and the whole of April and therefore as shown in Table 5.27, the lowest price of wheat is found either in April or in May. Borsad market yard had the lowest price in April and the price index was 91.79. On the other hand, Kapadvanj and Mehmedabad market yards experienced the lowest price in May and their respective price indices were 89.65 and 90.00. However, as an exception, Khambhat market yard has experienced its lowest price somewhat late, i.e. in July and its price index was 94.39. The highest price of rice is experienced by all the markets except Borsad market yard in the month of December. Borsad market yard has experienced the lowest price in the month of January. The price indices in Borsad, Khambhat, Kapadvanj and Mehmedabad market yards were 110.29, 104.60, 109.04 and 110.45 respectively. If we look at the graphs given in figures 9, 10, 11 and 12 (Appendix II) we find much uniform variations in the wheat prices in two market yards, viz. Borsad and Mehmedabad.

Bajra is a short-duration kharif crop and normally it begins to come to the market around October. So, the lowest price of bajra is found in October in all the four market yards. The price indices (as shown in Table 5.28) in Borsad, Khambhat, Kapadvanj and Mehmedabad market yards are 89.38, 89.12, 90.78 and 90.07 respectively. Again, there is no uniform month so far as the highest price of bajra is concerned but in general they are highest between February and April. In Mehmedabad and Khambhat market yards the highest price indices are found in the month of March and their values are 106.7 and 105.58 respectively. In Borsad market yard, the highest price index (107.66) is found in the month of April and in Kapadvanj market yard, it (105.99) is found in the month of February. Thus, there is no uniformity of month for the highest price of bajra. It can be seen from figures 13, 14, 15 and 16 (Appendix II) that there is almost the same pattern of the behaviour of prices in Borsad and Khambhat market yard. Moreover, the pattern of behaviour of prices in Mehmedabad and Kapadvanj market yard are similar to that of Borsad and Khambhat market yard but the extent and month of price variations do not coincide.

The above mentioned discussion of seasonal fluctuations in prices show that prices are not relatively stable over time, i.e. during the different seasons of the year. Thus, the hypothesis:

"Market regulation provides stability of market in terms of prices" is not found valid.

In the absence of stability as shown above, it would be worth studying whether the farmer really benefits by storing his produce at the time of harvest or at the time of low price and then selling the same at the time of high price. The farmer who stocks his produce loses by way of drainage, pilferage etc. Further, he loses income by way of interest on the value of his produce, if he stores in the present and sells in the future. To calculate the farmer's net gain or loss in this type of operation, it is necessary to know the average time span between the highest and the lowest price and the average rise in price in percentage terms. This information is derived in Tables 5.29 to 5.32 through the data presented in those tables. Then bank-lending rate of interest (18%) has been taken into consideration to work out the net gain or loss of the farmer.

The average time span between the highest and the lowest price is given in Tables 5.29 to 5.32. In Borsad market yard. the average time span between the highest and the lowest price for paddy, rice, wheat and bajra are 7.5, 5.6, 3.7 and 7.1 months respectively. The average price rise in percentage terms for paddy, rice, wheat and bajra are 40.3, 32.7, 35.5 and 39.5 respectively. By dividing the average rise in price in percentage terms by the average time span, we get price rise per month in percentage terms. Thus, these figures (which are not shown in the table) turn out to be 5.37, 5.84, 9.59 and 5.56 for paddy, rice, wheat and bajra respectively. The bank lending rate being 18%, the monthly interest cost (of storing the produce worth Rs.100) turns out to be Rs. 1.5. It means that if the farmer does not sell at the time of lowest price and sells at the time of highest price he loses Rs.1.50 by way of interest and gains Rs.5.37, Rs. 5.84, Rs. 9.59 and Rs. 5.60 by way of price rise in a sale worth Rs.100. Thus, the farmer has a net gain if he stores his produce and sells later at a higher price in Borsad market yard. This net gain for paddy, rice, wheat, and bajra is Rs. 3.87, Rs. 4.33, Rs.8.09 and Rs.4.06 in a sale of the produce worth Rs. 100.

In Khambhat market yard, the average time span between the highest and lowest price is given in Table 5.30. The average time span between the highest price and the lowest price for paddy, rice, wheat, and bajra are 6.3, 5.2, 5.3 and 6.9 respectively. The average price rise in percentage terms for paddy, rice, wheat and bajra are 36.3, 36.8, 32.0 and 39.1 respectively. By dividing the average rise in price in percentage terms by the average time span we get price rise per month in percentage terms. Thus, these figures (which are not shown in the table) turn out to be 5.76, 7.08, 6.04 and 5.67 respectively. The bank lending rate being 18% the monthly interest cost (of storing the produce worth Rs. 100) turns out to be Rs. 1.50. It means that if the farmer does not sell at the time of lowest price and sells at the time of highest price, he loses Rs.1.50 by way of interest and gains Rs. 5.76, Rs. 7.08, Rs.6.04 and Rs.5.67 by way of price rise in a sale worth Rs.100. Thus, the farmer has a net gain if he stores his produce and sells later at a higher price in Khambhat market yard. This net gain for paddy, rice, wheat and bajra is Rs.4.26, Rs.5.58, Rs.4.54 and Rs.4.17 in a sale of the produce worth Rs.100.

TABLE 5.29: Seasonal Movement of Paddy, Rice, Wheat and Bajra Prices During 1975-76 to 1984-85

	Borsad			
	Paddy	Rice	Wheat	Bajra
1. Range of Months between the Lowest and the Highest Price during 1975-76 to 1984-85	2 to 10	0 to 10	2 to 8	4 to 10
2. Average of Number of Months (i.e. time span) between the Lowest and the Highest Price during 1975-76 to 1984-85	7.5	5.6	3.7	7.1
3. Range of %age increase in Seasonal Price	14 to 64	2 to 63	17 to 80	19 to 69
4. Average of %age Increase in Seasonal Price	40.3	32.7	35.5	39.5

Note: Method of the derivation of the figures given in this table from original data has been given in Appendix III. Derivation of figures given in tables 5.29 to 5.32 is identical.

TABLE 5.30: Seasonal Movement of Paddy, Rice, Wheat and Bajra Prices During 1975-76 to 1984-85

	Khambhat			
	Paddy	Rice	Wheat	Bajra
1. Range of Months between the Lowest and the Highest Price during 1975-76 to 1984-85	2 to 9	2 to 10	1 to 10	4 to 10
2. Average of Number of Months (i.e. time span) between the Lowest and the Highest Price during 1975-76 to 1984-85	6.3	5.2	5.3	6.9
3. Range of %age increase in Seasonal Price	12 to 65	13 to 76	13 to 64	12 to 73
4. Average of %age Increase in Seasonal Price	36.3	36.8	32.0	39.1

TABLE 5.31: Seasonal Movement of Paddy, Rice, Wheat and Bajra Prices During 1975-76 to 1984-85

	Kapadvanj			
	Paddy	Rice	Wheat	Bajra
1. Range of Months between the Lowest and the Highest Price during 1975-76 to 1984-85	2 to 10	0 to 9	2 to 10	0 to 10
2. Average of Number of Months (i.e. time span) between the Lowest and the Highest Price during 1975-76 to 1984-85	7.1	6.5	5.4	4.8
3. Range of %age increase in Seasonal Price	19 to 100	18 to 56	23 to 78	21 to 84
4. Average of %age Increase in Seasonal Price	44.3	30.3	38.1	47.5

TABLE 5.32: Seasonal Movement of Paddy, Rice, Wheat and Bajra Prices During 1975-76 to 1984-85

	Mehmedabad			
	Paddy	Rice	Wheat	Bajra
1. Range of Months between the Lowest and the Highest Price during 1975-76 to 1984-85	1 to 10	0 to 10	1 to 10	4 to 10
2. Average of Number of Months (i.e. time span) between the Lowest and the Highest Price during 1975-76 to 1984-85	6.3	5.7	4.1	7.3
3. Range of %age increase in Seasonal Price	15 to 52	9 to 44	15 to 70	20 to 89
4. Average of %age Increase in Seasonal Price	35.0	22.4	34.2	41.0

In Kapadvanj market yard, the average time span between the highest and the lowest price is given in table 5.31. The average time span between the highest price and the lowest price for paddy, rice, wheat and bajra are 7.1, 6.5, 5.4 and 4.8 respectively. The average price rise in percentage terms for paddy, rice, wheat and bajra are 44.3, 30.3, 38.1 and 47.5 respectively. By dividing the average rise in price in percentage terms by the average time span we get price rise per month in percentage terms. Thus, these figures (which are not shown in table) turn out to be 6.24, 4.66, 7.05 and 9.90 respectively. The bank lending rate being 18%, the monthly interest cost (of storing the produce worth Rs.100) turns out to be Rs.1.50. It means that if the farmer does not sell at the time of lowest price he loses Rs.1.50 by way of interest and gains Rs.6.24, Rs.4.66, Rs.7.05, and Rs.9.90 by way of price rise in a sale worth Rs.100. Thus, the farmer has a net gain if he stores his produce and sells later at a higher price in Kapadvanj market. This net gain for paddy, rice, wheat, and bajra is Rs.4.74, Rs.3.16, Rs.5.55 and Rs.8.39 in a sale of produce worth Rs. 100.

In Mehmedabad market yard, the average time span between the highest and the lowest price is given in the Table 5.32. The average time span between the highest price and the lowest price for paddy, rice, wheat, and bajra are 6.3, 5.7, 4.1 and 7.3 respectively. The average price rise in percentage terms for paddy, rice, wheat and bajra are 35.0, 22.4, 34.2 and 41.0 respectively. By dividing the average rise in price in percentage terms by the average time span, we get price rise per month in percentage terms. Thus, these figures turn out to be 5.6, 3.93, 8.34 and 5.62 respectively. The bank lending rate being 18% the monthly interest cost (of storing the produce worth Rs.100) turns out to be Rs. 1.50. It means that if the farmer does not sell at the time of lowest price, he loses Rs. 1.50 by way of interest and gains Rs.5.6, Rs.3.93, Rs.8.34 and Rs.5.62 by way of price rise in a sale worth Rs.100. Thus, the farmer has a net gain if he stores his produce and sells later at a higher price in Mehmedabad market. This net gain for paddy, rice, wheat and bajra is Rs.4.06, Rs.2.43, Rs.6.84 and Rs.4.12 in a sale of produce worth Rs.100.

Thus, it may be concluded that the markets are not stable seasonally although they are spatially integrated. It means that they do not satisfy one of the conditions of the ideal market. They tend to encourage speculation. Finally, the hypothesis:

> "Market regulation provides stability of markets in terms of prices" is not proved in the study of selected markets of Kheda District.

Section IV

COST OF BUYING AND COST OF SELLING

Marketing cost is an important indicator of marketing efficiency. Before regulation was introduced in India, there were exorbitant market charges, unauthorised deductions, incorrect and false weights, superfluous middlemen, malpractices etc. Regulation of market removes all these defects of agricultural marketing. Thus, improvement in the marketing system brings reduction in marketing costs.

The range of maximum and minimum market charges prevailing in market yards are fixed by the government at the state level and the actual market charges accepted by the market committees are published in their annual reports. So, the information about market charges prevailing in regulated markets are collected from the annual reports of the markets. However, a primary survey of farmers and traders was conducted to find out the cost of buying and cost of selling inside and outside the yard. For this purpose, 90 farmers and 10 traders from each market area under regulation were surveyed in the year 1989 and the data pertaining to year 1987-88 was collected with the help of structured questionnaire. Thus, the total traders and farmers of the sample come down to 40 and 360 respectively. Traders' purchases inside and outside the yard are given in Table 5.33 to 5.35. The discussion in these tables is a cropwise discussion.

Traders' Purchases Inside and Outside the Market Yard

(i) *Paddy*

The annual requirement of paddy of 10 sample traders in the year 1987-88 in Borsad market was 29,500 quintals as shown in Table 5.33. Out of this, they purchased 17,500 quintals from the yard itself and 12,000 quintals from outside the yard. In the regulated market yard, out of 17,500 quintals, they purchased 14,000 quintals directly and 3,500 quintals through commission agents. So far as their purchase from outside the yard is concerned, out of 12,000 quintals, they purchased 42,000 quintals directly from the farmers, 3,600 quintals from other traders and 4,200 quintals through commission agents.

As shown in the table mentioned above in 1987-88 the annual requirements of paddy of 10 sample traders in Khambhat market area was 427,000 quintals. Out of this, they purchased 181,000 quintals from regulated market yard and 246,000 quintals from outside regulated market yard. In the yard, out of 181,000 quintals purchased, 87,000 quintals were purchased directly and 94,000 quintals through commission agents. The traders purchased 73,000 quintals from farmers, 12,000 quintals from other traders and 161,000 quintals through commission agents, so far as their purchases outside the yard were concerned.

In Mehmedabad market area, the annual requirements of paddy of 10 sample traders in the year 1987-88 was 145,000 quintals. Out of this, they purchased 116,500 quintals from the yard and 28,500 quintals from outside the regulated market yard. In the regulated market yard, 116,500 quintals were purchased directly. So far as the purchase outside the yard is concerned, out of 28,500 quintals, they purchased 26,500 quintals from the farmers and 2,000 quintals from other traders.

TABLE 5.33: Traders' Purchases Inside and Outside the Yard

PADDY *(In Quintals)*

Name of Market	Inside the Yard			Outside the Yard				Total (4+8)
	Direct	Commission Agent	Total (2+3)	From Farmer	Other Traders	Commission Agents	Total (5+6+7)	
(1)	(2)	(3)	(4)	(5)	(6)	(7)	(8)	(9)
Borsad	14000	3500	17500	4200	3600	4200	12000	29500
Khambhat	87000	94000	181000	73000	12000	161000	246000	42700
Kapadvanj	-	-	-	-	-	-	-	-
Mehmedabad	116500	-	116500	26500	2000	-	28500	145000

(ii) *Wheat*

The annual requirements of wheat of 10 sample traders in the year 1987-88 in Borsad market area was 31,700 quintals as shown in Table 5.34. Out of this, they purchased 25,200 quintals from regulated market yard and 6,500 quintals from outside the yard. In the regulated market yard, out of 25,200 quintals they purchased 21,000 quintals directly and 4,200 quintals through commission agents. So far as their purchase from outside the yard is concerned, out of 6,500 quintals, they purchased 3,600 quintals directly from the farmers, 300 quintals from other traders and 2,600 quintals through commission agents.

In Khambhat market area, the requirement of wheat of 10 sample traders in 1987-88 was 31,000 quintals. From regulated market yard, they purchased 23,000 quintals. Out of this, they purchased 18,000 quintals directly from the regulated market and 5,000 quintals through the commission agents. They purchased 8,000 quintals of their requirements from outside the yard and out of this, 2,000 quintals from the farmers and 6,000 quintals through the commission agents.

The annual requirement of wheat of the 10 sample traders of Kapadvanj market area in 1987-88 was 6,500 quintals. Out of this 6,500 quintals, they purchased 5,600 quintals directly from the regulated market yard and the remaining requirements, i.e. 900 quintals were purchased from outside the yard. Out of this 900 quintals, they purchased 500 quintals from the farmers and 400 quintals through commission agents.

In Mehmedabad market area, the annual requirements of wheat of the 10 sample traders in 1987-88 was 52,850 quintals. Out of this, they purchased 35,850 quintals from the regulated market yard and the remaining 17,000 quintals from outside the yard. In the regulated market yard, the traders purchased 35,850 quintals directly. Outside the market yard, 17,000 quintals were purchased directly from the farmers.

(iii) *Bajra*

So far as bajra is concerned, the annual requirement of the 10 sample traders of Borsad market area in 1987-88 was 43,500 quintals as shown in Table 5.35. Out of this, 33,500 quintals were purchased from the regulated market yard and the remaining 10,000 quintals from outside the yard. In the regulated market yard, the traders purchased 29,000 quintals directly and 4,500 quintals through commission agents.

TABLE 5.34: Traders' Purchases Inside and Outside the Yard

(*In Quintals*)

Name of Market	WHEAT							
	Inside the Yard			Outside the Yard				Total (4+8)
	Direct	Commission Agent	Total (2+3)	From Farmer	Other Traders	Commission Agents	Total (5+6+7)	
(1)	(2)	(3)	(4)	(5)	(6)	(7)	(8)	(9)
Borsad	21000	4200	25200	3600	300	2600	6500	31700
Khambhat	18000	5000	23000	2000	-	6000	8000	31000
Kapadvanj	5600	-	5600	500	-	400	900	6500
Mehmedabad	35850	-	35850	17000	-	-	17000	52850

Outside the market yard, they purchased 6,800 quintals from the farmers and 3,200 quintals through commission agents.

In Khambhat market area, the annual requirement of bajra of the 10 sample traders was 12,500 quintals. Out of 12,500 quintals, 10,200 quintals were purchased from the regulated market yard and 2,300 quintals from outside the yard. In the regulated market yard, 7,000 quintals were purchased directly and 3,200 quintals through the commission agents. Outside the yard, the traders purchased 1,250 quintals from the farmers, 150 quintals from other traders, 900 quintals through the commission agents.

The 10 sample traders of Kapadvanj market area in 1987-88 required 9,000 quintals of bajra, out of which they bought 7,900 quintals from regulated market yard and 1,100 quintals from outside the yard. In the regulated market yard, the traders bought 7,900 quintals directly. Outside the yard, the traders purchased 500 quintals from the farmers and 600 quintals through the commission agents.

In Mehmedabad market area, the 10 sample traders required 41,750 quintals of bajra in 1987-88, out of which they purchased 28,250 quintals from the regulated market yard and 13,500 quintals from outside the yard. In the regulated market yard, the traders purchased directly and outside the yard also they purchased directly from the farmers. There was no commission agent involved in the transaction in Mehmedabad market area so far as purchase of bajra by the sample traders is concerned.

Traders' Cost of Buying and Selling Inside and Outside the Market Yard

Traders' cost of buying inside and outside the regulated market yard is presented in Table 5.36. It can be seen from the table that so far as commission charges are concerned, traders benefit when they purchase from outside the market yard. The only exception is Mehmedabad, where commission charges outside market yard (Rs.1.50) exceed the commission charges (Rs.0.75) payable in regulated market yard. In case of Tolai, it is more outside the yard than what it is in regulated market yard. This is true for all the selected areas viz. Borsad, Khambhat, Kapadvanj and Mehmedabad. It is found from the sample study of traders that no Hamali is to be paid outside the yard whereas Hamali is to be paid at a fixed rate inside the yard. Perhaps, either the tolat or the farmer himself does this work outside the yard. The rates of

TABLE 5.35: Traders' Purchases Inside and Outside the Yard

(In Quintals)

	BAJRA							
	Inside the Yard			Outside the Yard				
Name of Market	Direct	Commission Agent	Total (2+3)	From Farmer	Other Traders	Commission Agents	Total (5+6+7)	Total (4+8)
(1)	(2)	(3)	(4)	(5)	(6)	(7)	(8)	(9)
Borsad	29000	4500	33500	6000	-	3200	10000	43500
Khambhat	7000	3200	10200	1250	150	900	2300	12500
Kapadvanj	7900	-	7900	500	-	600	1100	9000
Mehmedabad	28250	-	28250	13500	-	-	13500	41750

TABLE 5.36: Market Charges Inside and Outside Regulated Market Yards

(*In Rupees*)

Name of the Market	Change Inside the Yard				Charges Outside the Yard			
	Commission Charges per Rs.100	Tolai per bag	Hamali per bag	Market fee per Rs.100	Commission Charges per Rs.100	Tolai per bag	Hamali per bag Rs.100	Market fee per Rs.100
Borsad	0.75	0.04	0.10	0.30	0.40	1.00	-	0.30
Khambhat	0.70	0.10	0.30	0.30	0.40	1.00	-	0.30
Kapadvanj	0.50	0.15	0.07	0.30	0.40	1.00	-	0.30
Mehmedabad	0.75	0.15	0.05	0.30	1.50	1.00	-	0.30

Hamali differ in different market as shown in the Table 5.36. Market fee is uniform i.e. 30 paise per Rs.100 in both inside and outside the yard. On the whole, it is advantageous to the trader to purchase from a yard because absence of Hamali and low commission charges outside the yard are over compensated by excessive Tolai charges there e.g. in Borsad market yard for a given value of Rs.100 for the produce contained in the bag, the trader will pay Rs.1.19 and Rs.1.70 in the yard and outside the yard respectively by way of buying charges. To repeat, traders' cost of buying charges are less in the yard as compared to those payable outside the yard. This is true for all the selected market yards, viz. Borsad, Khambhat, Kapadvanj and Mehmedabad.

Coming to the cost of selling, transport cost is the only cost borne by the farmer. This cost is to be paid by him to bring his produce from his village to the market place. All other charges like market fee, commission charges, Tolai, hamali etc. are to be paid by the trader. Table 5.37 presents the transport cost paid by the farmer in different market areas. This information is obtained from the sample study of the farmers. The data presented in Table 5.37 shows the transport cost paid by the representative farmer of each size group,i.e. marginal, small and large inside and outside the yards of Borsad, Khambhat, Kapadvanj, and Mehmedabad market areas. It can be seen from the table that transport cost per quintal borne by the large farmer is less as compared to that borne by the marginal and small farmers while selling in the yard with the exception of Mehmedabad for villages within five kilometres and Kapadvanj for village within fifteen kilometres. The reason for this difference might be due to the fact that the marginal and the small farmer have a small quantity to be transported to the yard and on the other hand, the transportation charges might be some fixed minimum per trip. Alternatively speaking, the trip of a lorry/cart has some excess capacity and therefore, per quintal transport cost of a marginal and small farmer is higher than that paid by a large farmer because the large farmer has enough quantity to be transported in one trip of a lorry/cart. As a result, per quintal transport cost of a large farmer is likely to be low. It appears from the table that transportation charges are relatively low in Borsad market area and high in Kapadvanj market area. The reasons for this difference cannot be known from the available data of the sample study. Further, so far as transport cost for the sale in market yard and outside yard is concerned, it is found that the transport cost in each size group of each market yard (except Khambhat market area where information is not available) is relatively less for the sale outside the yard as compared to the sale inside the yard. The reason for this difference might be due to the fact that the distance of a market yard might be more than the distance of a trading place outside the yard.

TABLE 5.37: Transport Cost per Quintal According to Size Group of Holding and According to Distance from Market Yard

(In Rupees)

Size Group	Transport Cost per quintal inside the market yard		Transport cost per quintal outside the market yard
	Villages within 5 kms	Villages within 15 kms	
(1)	(2)	(3)	(4)
Borsad			
1. Marginal	3.14	**	*
2. Small	2.54	3.60	0.90
3. Large	1.97	2.75	0.60
Khambhat			
1. Marginal	3.18	4.92	*
2. Small	2.88	4.94	*
3. Large	2.74	4.74	*
Kapadvanj			
1. Marginal	3.90	5.00	2.44
2. Small	4.27	5.00	2.81
3. Large	3.39	5.00	2.17
Mehmedabad			
1. Marginal	3.38	3.60	1.16
2. Small	3.31	4.32	1.24
3. Large	3.80	4.17	1.60

* Information is not available because the farmers surveyed happened to be the sellers selling exclusively in regulated market yard.

** The farmers coming to Borsad market yard from a village 15 kms away from the yard could not be identified and therefore information was not available.

Besides transport cost, one more dimension of marketing cost is weighment. It is found from the farmers surveyed that scales and weights are alright both in inside the yard and outside the yard. Farmers believe that they are not cheated outside the yard by way of defective scales and false weights. However, they agree to the fact that they have to give 1 kg. of produce more per quintal when they sell outside the yard. The value of this 1 kg. should be added in the cost of selling of a farmer for his sale outside the yard. The value of this 1 kg. might vary from crop to crop and from time to time. Therefore, there cannot be any generalisation about this. However, at current prices of paddy, wheat and bajra in the year 1988-89, if price of 1 kg. for each size group is added to the cost of transport, farmer's total cost of selling outside the market yard exceeds that of inside the market yards.

Thus, the hypothesis,

"The users of the regulated market yards enjoy cost economies".

is found valid for both farmers and traders from the point of view of cost of selling and cost of buying.

When the traders were asked the reasons for purchasing from regulated markets, they gave the following reasons which were common for all the traders in all the selected markets:

(i) Low transport cost.
(ii) Supply available from nearby markets.
(iii) Financial ease or facility.
(iv) Licence holding.
(v) Possibility of verification of the quality of the produce before purchase.
(vi) Scope for the choice of produce (because there are many sellers).
(vii) Reasonable prices.
(viii) Proper weights and measures.
(ix) Certainty of venue for the purchase.
(x) Availability of godown for storage.
(xi) Sufficient quantity is available.
(xii) Easy purchase due to auctioning.

Traders have expressed the following difficulties when they purchase from a regulated market yard:

(1) There are more purchasers for a given supply and therefore they have to purchase in small quantities.
(2) Extent of competitiveness is more and therefore they have to pay a higher price.
(3) Most of the produce come through commission agents and commission agents are inadequate in number. Further, they do not allow the produce to come to the market yard.
(4) They have to make cash payment to the commission agent on the third day.
(5) Auctioning is not healthy.
(6) Stock-register has to be furnished on every Friday.
(7) All types of produce, i.e. good quality produce does not come to the market yard.
(8) Produce is not always homogeneous.
(9) Market staff is less cooperative.

Producers' Price and Sale Inside and Outside Market Yard

Based on primary survey, Table 5.38 give crop-wise producer's price for each size group inside and outside market yard of the areas selected for the study.

It can be seen from the Table 5.38 that the prices received by marginal, small and large farmers for Bajra in Borsad market yard are Rs.164.0, Rs.164.70 and Rs.165.65 respectively.

Similarly, in Khambhat market yard, for wheat the prices received by the marginal, small and large farmers are Rs.280.65, Rs.292.38 and Rs.296.92 respectively. This pattern of prices shows that the marginal farmer gets a lower price than that received by the small farmer and the price received by the small farmer is less than the price received by the large farmer. The same pattern is observed for the sale of paddy outside the regulated market yard in Borsad and Mehmedabad market areas. Elsewhere, the same pattern is not found in the systematic form. However, as a part of the objectives of regulating the markets, it would be desirable that the marginal and the small farmers do not get price less than that got by the large farmers. The probable reason for fetching low price on the part of marginal and small farmers may be the compulsion to sell early to meet the urgent need for finance. If this is so, some arrangement should be made so that their financial needs can be satisfied and they are enabled to sell at a time when they get a better price. This is a policy problem which will be elaborated in the last chapter while discussing policy prescriptions.

TABLE 5.38: Producers' Price Inside and Outside the Regulated Market Yards for Farmers of Different Size Groups

(*Rs. per quintal*)

Market Area	Size Group	Inside Yard			Outside Yard		
		Paddy	Wheat	Bajra	Paddy	Wheat	Bajra
Borsad	1. Marginal	230.45	280.00	164.00	210.00	210.00	151.67
	2. Small	226.00	250.45	164.70	237.50	245.00	158.00
	3. Large	228.65	265.56	165.65	—	253.57	176.00
Khambhat	1. Marginal	—	280.65	161.85	—	—	—
	2. Small	—	292.38	161.29	—	—	—
	3. Large	—	296.92	159.79	—	—	—
Kapadvanj	1.Marginal	—	242.75	156.67	—	245.00	181.00
	2.Small	—	257.53	164.89	—	200.00	161.00
	3.Large	—	244.09	151.54	—	213.18	156.67
Mehmedabad	1.Marginal	228.21	229.00	160.00	230.00	245.00	167.00
	2.Small	226.78	232.70	159.44	-	220.00	260.00
	3.Large	252.16	222.90	162.95	250.00	222.55	159.58

TABLE 5.39: Sale Inside and Outside the Regulated Market Yard According to Size Group of Holding and Crop in Selected Markets

(In percentage)

Market	Size Group	Paddy		Wheat		Bajra	
		Sale inside the yard	Sale outside the yard	Sale inside the yard	Sale outside the yard	Sale inside the yard	Sale outside the yard
Borsad	Marginal	90.90	9.10	81.43	5.71	71.33	28.67
	Small	75.15	23.08	69.07	24.74	76.92	23.08
	Large	100.00	—	75.60	23.59	80.70	19.30
Khambhat	Marginal	—	—	89.94	—	93.18	—
	Small	—	—	98.26	—	98.40	—
	Large	—	—	100.00	—	95.38	—
Kapadvanj	Marginal	—	—	79.74	14.38	72.56	27.43
	Small	—	—	84.50	14.75	79.31	20.69
	Large	—	—	85.60	14.40	79.49	20.51
Mehmedabad	Marginal	97.00	3.00	55.04	44.96	70.00	28.00
	Small	89.31	—	77.36	22.64	52.97	47.03
	Large	98.91	1.09	82.70	17.08	61.56	38.44

Table 5.39 shows producers' sale as percentage of their marketable surplus inside the yard and outside the yard. In some cases the total of the two does not reach 100. The reason may be that they might have stored a part of their surplus to sell afterwards.

If price for each crop in each size group is compared for each area then the emerging picture is like this. In Borsad market area, barring small paddy growers and large bajra growers, all other growers get a lower price per quintal outside the yard as shown in Table 5.38. (In Khambhat market area, all the surveyed farmers sold their produce in the market yard as shown in Table 5.39). Therefore, prices outside market yard are not available and comparison is not possible. In Kapadvanj market area, except large wheat growers, all get lower prices inside the yard than those got outside the yard. In Mehmedabad market area marginal growers of paddy, wheat and bajra and the small growers of bajra get a lower price inside the yard than that got outside the yard.

From the above mentioned factual position, it can be said that the farmers of Borsad market area should be advised to sell the whole of their surplus in the regulated market yard only because by doing so they can get a better price for their produce and can maximize their incomes. It can be seen from Table 5.39 that the farmers of Borsad market area do sell outside the market yard and therefore, there is a definite scope for improving their economic conditions. In Kapadvanj and Mehmedabad market areas, the farmers get a higher price in general outside the regulated market yard. This means that these regulated market yards are not sufficiently competitive and therefore steps should be taken to make them more competitive and more efficient, so that the purpose of establishing the regulated market yards is fully served. After that the farmers of this market should be advised to sell their surplus in the regulated market yards only. Further, it is obvious from the above discussion that comparatively, yard of Borsad market is more efficient than yards of Kapadvanj and Mehmedabad.

Section V

MARKET ARRIVALS

Market arrivals can show the extent of regulated market consciousness among the producers. Market arrival is an index of the producer's willingness to participate in the regulated market. Joshi[12] is of the opinion that the efficiency of regulated market can be tested by the extent of the participation by the farmers in bringing their produce to the market for sale. The market arrival is one of the performance variables of the regulated market.

The efficiency of the regulated market can be judged by the volume of the agricultural produce that the regulated market can attract.

Table 5.40 shows the market arrival of the notified crops during the period 1975-76 to 1984-85 in Kheda district. During the study period, the arrivals of the notified crops was maximum in the year 1979-80. Ever since that year, it has started declining but it picked up again in 1983-84.

Table 5.41 shows the percentage distribution of the selected notified crops to the market arrivals during the period. Paddy, wheat, and bajra are the important crops arriving in the regulated market of Kheda district. Nearly 50-60 per cent of the market arrivals comprised of the above crops in almost all the years of the study period. In fact, in the year 1982-83, this percentage was even higher, i.e. nearly 80.76 per cent of arrivals consisted of paddy, wheat and bajra.

TABLE 5.40: Market Arrivals of the Notified Crops During the Period 1975-76 to 1984-85 in Kheda District

(In lakh quintals)

Year	Volume of arrivals	Increase (+) or Decrease (-) over previous year
1975-76	40.81	—
1976-77	35.37	- 5.43
1977-78	36.62	+ 1.24
1978-79	49.29	+ 12.67
1979-80	91.39	+ 42.10
1980-81	41.39	- 50.00
1981-82	44.95	+ 3.56
1982-83	41.94	- 3.01
1983-84	58.53	+ 16.59
1984-85	50.89	- 7.64

Source: Annual Report of the District.

Out of the three crops viz. paddy, wheat and bajra paddy dominates the scene of market arrival in the regulated market yards.

Market concentration of paddy, wheat, and bajra in selected markets are shown in Table 5.42 to 5.44. So far as paddy is concerned, Khambhat market yard dominates in terms of percentage of paddy arrival followed by Mehmedabad, Borsad and Kapadvanj market yard during the reference period. In case of wheat, out of the four regulated market yards, Mehmedabad market yard attracts relatively more arrival in percentage terms followed by Borsad, Khambhat and Kapadvanj market yards during 1975-76 to 1984-85. In case of bajra, during the study period, Borsad market yard received more arrival in percentage terms followed by Mehmedabad, Khambhat and Kapadvanj market yards.

The average market arrival per market during the study period is given in Table 5.45. The average market arrival has increased over years. In 1975-76, the average market arrival was 2.09 lakh quintals and this average increased to 3.24 lakh quintals in 1984-85. The average market arrival per market in the district as a whole was 2.98 lakh quintals during the period 1975-76 to 1984-85.

TABLE 5.41: Percentage Distribution of the Market Arrivals of the Selected Crops to the Total Arrivals During the Period 1975-76 to 1984-85

Year	Paddy	Wheat	Bajra	Others	Total
1975-76	36.44	12.19	7.93	43.44	100.00
1976-77	42.55	12.08	9.26	36.11	100.00
1977-78	49.85	11.42	10.72	28.01	100.00
1978-79	44.43	10.55	9.51	35.51	100.00
1979-80	36.59	5.38	5.13	52.90	100.00
1980-81	54.40	10.42	11.21	23.97	100.00
1981-82	52.70	9.07	12.04	26.19	100.00
1982-83	54.90	11.46	14.40	19.24	100.00
1983-84	62.46	9.90	8.52	19.12	100.00
1984-85	52.62	10.12	7.49	29.77	100.00

The market-wise data regarding the average market arrivals for the whole period is also given in Table 5.45. The market arrivals of a majority of the regulated markets are below the average. Taking the whole period into consideration, the number of market yards receiving arrivals below the market average is six. Thus, it may be said that 55 per cent of the regulated market yards in the Kheda district are working unsatisfactorily as their market arrivals for the study period are less than the district average of market arrivals in regulated market yards of Kheda.

TABLE 5.42: Market Concentration of Paddy Arrivals in Selected Market Yards During 1975-76 to 1984-85 (*In percentage to the total arrivals*)

Year	Borsad	Khambhat	Kapadvanj	Mehmedabad
1975-76	19.24	54.43	4.15	59.95
1976-77	22.93	58.34	8.61	64.14
1977-78	26.44	70.21	14.98	68.15
1978-79	13.21	63.66	13.55	63.64
1979-80	32.53	61.60	12.33	60.57
1980-81	33.73	72.09	16.95	64.52
1981-82	39.15	73.13	20.31	62.28
1982-83	33.79	72.17	11.70	58.44
1983-84	40.15	67.11	28.87	72.85
1984-85	38.82	83.35	14.49	67.82

TABLE 5.43: Market Concentration of Wheat Arrivals in Selected Market Yards During 1975-76 to 1984-85 (*In percentage to the total arrivals*)

Year	Borsad	Khambhat	Kapadvanj	Mehmedabad
1975-76	26.22	20.50	5.62	16.66
1976-77	9.63	18.22	3.19	19.41
1977-78	14.12	8.07	5.83	12.22
1978-79	7.29	12.52	4.62	15.56
1979-80	7.88	9.39	5.20	16.25
1980-81	9.96	6.35	5.66	12.40
1981-82	11.31	6.41	6.68	14.65
1982-83	14.86	7.20	5.57	18.00
1983-84	7.19	17.29	8.68	14.77
1984-85	10.09	4.89	4.91	20.05

TABLE 5.44: Market Concentration of Bajra Arrivals in Selected Market Yards During 1975-76 to 1984-85 (*In percentage to the total arrivals*)

Year	Borsad	Khambhat	Kapadvanj	Mehmedabad
1975-76	17.83	6.05	4.24	6.08
1976-77	26.12	7.08	5.56	7.22
1977-78	26.38	7.64	6.98	8.80
1978-79	15.87	8.04	8.60	9.56
1979-80	27.09	13.89	8.04	12.47
1980-81	29.60	8.44	8.60	12.75
1981-82	21.87	8.25	11.87	14.32
1982-83	38.67	11.01	9.49	17.44
1983-84	19.78	6.68	9.68	7.56
1984-85	16.85	4.29	5.13	9.36

TABLE 5.45: Arrival of Paddy, Wheat and Bajra to Regulated Market Yards During the Period 1975-76 to 1984-85

(*In lakh quintals*)

Name of the Regulated Market	1975-76	1976-77	1977-78	1978-79	1979-80	1980-81	1981-82	1982-83	1983-84	1984-85	Average for the whole period
Average quantity of market-arrival per market	2.09	2.05	2.39	2.89	3.91	2.86	3.01	3.07	4.30	3.24	2.98
1. Anand	1.13	1.51	1.64	1.45	1.25	1.74	1.79	2.15	1.39	0.97	1.50
2. Kapadvanj	0.70	1.02	1.17	1.60	1.45	1.43	1.69	1.27	2.01	1.49	1.38
3. Khambhat	3.77	2.79	4.63	4.63	4.25	5.69	7.57	7.54	8.97	9.12	5.90
4. Thasra	1.19	1.38	0.61	1.90	1.95	2.08	2.17	7.60	1.75	1.66	2.23
5. Nadiad	3.51	3.42	4.08	4.22	18.53	4.24	4.19	3.42	18.53	6.97	7.11
6. Petlad	0.06	0.09	0.10	0.20	0.22	0.21	0.24	0.25	0.28	0.30	0.20
7. Balasinor	0.05	0.03	0.21	0.06	0.07	0.05	0.04	0.03	0.03	0.81	0.14
8. Borsad	2.92	1.97	2.84	3.26	2.98	3.23	3.82	3.66	2.87	3.16	3.07
9. Mehmedabad	2.60	3.74	3.44	4.14	2.76	2.84	3.76	3.51	3.98	4.11	3.49
10. Matar	6.13	5.48	6.43	9.00	8.37	8.44	5.81	2.51	5.44	4.56	6.22
11. Umreth	0.93	1.11	1.17	1.27	1.17	1.46	2.05	1.87	2.04	2.52	1.57
Total 23.04	22.54	26.32	31.73	43.00	31.41	33.13	33.81	47.29	35.67		
No. of Less Efficient Markets in the year	6 (55)	7 (64)	6 (55)	6 (55)	8 (73)	7 (64)	6 (55)	6 (55)	8 (73)	7 (64)	6 (55)

Note: The figures in parenthesis show the percentage of less efficient markets to the total markets (i.e. eleven markets) of the district.

Table 5.45 shows the number of less efficient markets too (i.e. markets receiving arrivals below the district average) in each year of the reference period and their percentage to the total number of markets. Out of 11 regulated market yards, there are six regulated market yards which are less efficient. The six less efficient market yards in the district are Anand, Kapadvanj, Thasra, Petlad, Balasinor, and Umreth. The average market arrival in these market yards are less than 2.98 lakh quintals of paddy, wheat, and bajra.

Table 5.46 shows market arrivals as a percentage of marketable surplus (as defined in Chapter III). This table is based on the data collected through primary survey. It shows market arrivals as a percentage of marketable surplus for each size group and each selected crop in four selected markets.

On the basis of total arrivals of each crop, following conclusions are drawn:

In case of market arrivals of paddy, the performance of Mehmedabad market yard is better than that of Borsad.

So far as wheat is concerned, the performance in terms of market arrivals as a percentage of marketable surplus, the market yards can be arranged in the descending order like this: (1) Khambhat, (2) Mehmedabad, (3) Kapadvanj, and (4) Borsad.

With regard to Bajra, if the market yards are arranged in descending order of their arrival performance, the ranking can be like this: (1) Khambhat, (2) Kapadvanj, (3) Borsad and (4) Mehmedabad.

On the basis of above findings, suitable policy measures can be devised to improve the arrival performance keeping in view the size groups and crops in each market yard.

Reasons for Low Arrivals

An attempt was made through the primary survey of the farmers to find out the reasons for low arrivals in the regulated market yards. In the questionnaire, following reasons were put before the farmers and were asked to rank them according to their own personal belief and experience. The reasons are as under: (a) a large quantity can be sold without the trouble of getting gunnies for bagging; (b) to meet urgent financial needs; (c) to clear tied loans; (d) saving in time; (e) uneconomical to take it to regulated market yards as the quantity is very small; (f) dissatisfaction with the regulated market yards.

TABLE 5.46: Market Arrival as a Percentage of Marketable Surplus According to Size Group of Holding and Crops*

Size Group	Paddy				Wheat				Bajra			
	Borsad	Khambhat	Kapad-vanj	Mehmeda-bad	Borsad	Khambhat	Kapad-vanj	Mehmeda-bad	Borsad	Khambhat	Kapad-vanj	Mehmeda-bad
Marginal	90.90	-	-	97.00	81.43	89.94	79.74	97.00	71.33	93.18	72.56	70.00
Small	75.15	-	-	39.31	69.07	98.26	84.50	89.31	76.92	98.40	79.31	52.97
Large	100.00	-	-	98.91	75.60	100.00	85.60	98.91	80.70	95.38	79.49	61.56
Total	91.56	-	-	96.51	74.25	96.59	84.19	96.51	77.89	96.06	78.07	60.39

* Based on primary survey conducted during April, 1989.

In Borsad market area, it is found that out of 53 responding farmers, 31 farmers ranked reason (a) first and 16 farmers ranked reason (b) first. Out of 31 farmers who ranked (a) first, 26 farmers were marginal and small ones. Similarly, out of 16 farmers who ranked (b) first, 11 farmers were small and marginal ones.

The farmers in Khambhat market area have not given reasons for not selling in regulated market yards mainly because most of them sell in regulated market yards only.

In Kapadvanj market area, out of 48 responding farmers, 27 farmers ranked reason (a) first and 11 farmers ranked (b) first. Out of 27 farmers who ranked (a) first, 13 farmers were marginal and small ones. Similarly, out of 11 farmers who ranked (b) first, 7 farmers were marginal and small ones.

In Mehmedabad market area, out of 46 responding farmers, 11 farmers ranked reason (a) first and 14 farmers ranked (b) first. Out of 11 farmers who ranked reason (a) first, 7 farmers were small and marginal ones. Similarly, out of 14 farmers who ranked (b) first, 9 farmers were marginal and small ones.

From the above analysis, it can be inferred that reasons (a) and (b) viz. (a) a large quantity can be sold without the trouble of getting gunnies for bagging, (b) to meet the urgent financial needs are the most prominent ones in all market areas except Khambhat. Further, among the farmers of different size groups, a relatively large number of small and marginal farmers experience these difficulties. Dissatisfaction with the market yard is not the reason for selling outside the market yard for any farmer in any market area. In other words, there is not a single farmer in any of the selected market areas who is dissatisfied with the working of the regulated market yards.

Agriculture Policy Commission is of the opinion that the facility of regulated market yard should be available to the farmers within the radius of five kilometres. In other words, distance should not be a barrier to the farmers to take their produce to the regulated market yard and the market yard should be negotiable either by foot or by cart. Thus, to examine the impact of distance, in the primary survey of the three villages attached to each market area, two villages were selected within the radius of five kilometres and one village was selected at a distance of approximately fifteen kilometres from the market yard.

If one looks at the problem from the point of view of size groups of farmers, it is observed that the impact of distance is there on marginal farmers of Borsad and Kapadvanj market area in all the selected crops. In other words, the market arrival (as a percentage of marketable surplus) of the distant village is less than the marketable surplus of nearby villages. In Mehmedabad market area, distance affects the marginal wheat and bajra growers but the marginal paddy growers are not affected by the distance. Another important and noteworthy point is that in Borsad market area, the market arrival of paddy and wheat of marginal farmers of a distant village is zero whereas in Kapadvanj and Mehmedabad market areas, the market arrival of marginal farmers of a distant village is positive but in percentage terms, it is less than the market arrival of the marginal farmers of nearby villages. The distance affects the small and large growers of selected crops of the selected market except the large farmers of a distant village of Borsad market area. Jasdanwala[13] also finds in her study that distance is the most important factor in the field of agricultural marketing. In other words, the market arrivals as a percentage of marketable surplus of the small and large growers of distant villages are less than the market arrivals of small and large growers of nearby villages.

Non-economic Variables

The non-economic variables include grading, disposal quickness, delay in payment, amenities, helpfulness of the market staff and functionaries, corruption and crime, pilferages and wastes, market consciousness and market information.

Grading facility is not available for the grading of crops in the regulated market yards of Kheda district. It is desirable that grading is introduced for the benefit of the farmers.

So far as disposal quickness is concerned, the farmers have to wait for 2 to 3 hours during the peak season. It all depends on the number of farmers participating in the auction on a particular day. An attempt should be made to reduce it to the minimum as far as possible.

How quick the payment is made to the farmer after the sale of his produce is an indicator of the efficiency of the market. From the primary survey of the farmers, it was found that during peak season, they had to wait for 3 to 4 hours to get their payment. In some exceptional cases, they had to stay overnight to get their payment. This is a problem of administrative procedure. If there is a scope to reduce the waiting period, all efforts should be made in that direction.

TABLE 5.47: Number of Farmers Listening to Market Information on All India Radio According to Size Group of Holding and Villages in Selected Market Areas

Market/ Village	Marginal (1)		Small (2)		Large (3)	
	Yes	No	Yes	No	Yes	No
Borsad						
Village						
Davol	1	9	1	9	4	6
Nisarya	4	6	-	10	1	9
Napatalpad	2	8	-	10	-	10
Total	7	23	1	29	5	25
Khambhat						
Village						
Shakarpur	5	5	6	4	4	6
Metpur	8	2	2	8	3	7
Ralej	4	6	3	7	2	8
Total	17	13	11	19	9	21
Kapadvanj						
Village						
Navagam	1	9	-	10	-	10
Jaleya	1	9	2	8	8	2
Atarsumba	-	10	-	10	-	10
Total	2	28	2	28	8	22
Mehmedabad						
Village						
Khatraj	-	10	-	10	-	10
Kachchai	-	10	-	10	-	10
Modaj	-	10	-	10	-	10
Total	-	30	-	30	-	30

Note: The first two villages in each market area are within the radius of 5 kms. and the third one is within the radius of 15 kms. from the market yard.

Physical amenities available in the market yards are also important in attracting the farmers to bring their produce to the yard. These facilities include drinking water facilities, water cisteners for the cattle, urinals and latrines, rest room for farmers and traders, canteen, bank, etc. All these facilities are available in the market yards selected for the study. So, from the point of view of physical amenities, these market yards are well equipped.

So far as the helpfulness of the market staff to the farmers is concerned, the farmers reported at the time of their interview that the market staff is fully helpful and cooperative. However, traders report that they are not cooperative to them.

Coming to market consciousness and market information of the farmers, the farmers were asked during their interview whether they listen to All India Radio broadcast of prices prevailing in regulated market yards or not, the situation was not satisfactory. It is presented in the tabulated form in Table 5.47.

It can be seen from the Table 5.47 that for farmers,of all the size groups of all the market areas except Khambhat, the situation is deplorable in the sense that the majority of the farmers do not listen to All India Radio broadcast of prices prevailing in the regulated market yards. Of the four market areas, the situation of Mehmedabad farmers is the worst in the sense that there is not a single farmer out of surveyed farmers (90) who listens to this programme of All India Radio.

Coming to corruption and crime, it was reported by the surveyed farmers during their interview that they had not to bribe any of the market functionaries at any stage of marketing. This is a positive and bright side of the working of the selected regulated markets.

In case of pilferages and wastes, the interviewed farmers reported that there are no major pilferages and wastes in the regulated market yards.

Thus, except the absence of grading and market consciousness as well as information, the market yards are efficient from the point of view of working of non-economic variables.

Conclusion

(1) Market yards (selected in the present study) are found to be competitive in terms of number of buyers and sellers, buyer concentration, freedom of entry-exit and mobility.

(2) Markets are spatially integrated.

(3) All the selected market yards are not equally stable in terms of prices. Borsad market yard is the most unstable and Mehmedabad is the most stable one. If they are ranked in descending order, they are - Mehmedabad, Kapadvanj, Khambhat and Borsad.

(4) Cyclical variations in prices around the trend for each crop in each market differ from one another, i.e. the intensity and the length of a cycle differ from case to case.

(5) In case of cost of buying and selling, both the traders and the farmers experience cost economies if they purchase and sell in regulated market yards.

(6) So far as the prices received by the farmers belonging to different size groups for different crops in different markets are concerned, no clear picture emerges from the data (see Table 5.38).

(7) In Khambhat, Nadiad, Borsad, Mehmedabad and Matar market yards, the market arrival is above the district average arrival and the rest have arrival below the district average. (This conclusion is based on secondary data for a period of 10 years.)

(8) Market efficiency measured in terms of market arrivals as a percentage of marketable surplus shows that markets can be arranged as under:

Paddy - Mehmedabad, Borsad
Wheat - Khambhat, Mehmedabad, Kapadvanj and Borsad.
Bajra - Khambhat, Kapadvanj, Borsad and Mehmedabad.

(9) Distance affects the farmers of all size group in all the selected market areas. In other words, market arrival as a percentage of marketable surplus of marginal, small and large farmers of nearby villages is more than that of distant villages.

(10) There are no complaints on the part of farmers with respect to disposal quickness, delay in payment, physical amenities, helpfulness of the market staff, corruption and crime, pilferages and wastes.

(11) Grading is absent in all the market yards.

(12) There is lack of market information and market consciousness on the part of farmers.

Along with the above findings, following hypotheses are tested:

(1) "Regulated markets have the features and characteristics of a competitive market" is found valid in the study of Kheda district.
(2) "Market regulation creates integration" is proved valid in the study of Kheda district.
(3) "Market regulation provides stability of market in terms of prices" is not proved valid in the study of Kheda district.
(4) "The users of the regulated market yards enjoy cost economies" is found valid for both the farmers and the traders from the point of view of cost of selling and cost of buying.

NOTES

1. Z.Y. Jasdanwalla, *op.cit.*
2. J.C. Abbot, "The Role of Marketing in the Development of Backward Agricultural Economics", **Journal of Farm Economics,** 11(4) May 1962, pp.349-352.
3. P.N. Kaliyaperumal, 'The Organisation Working and Performance of Regulated Markets in South Arcot District" (Unpublished Ph.D. Thesis, Madurai Kamaraj University), 1981.
4. If the same trader has purchased again on some other date, it is presumed that he is an independent trader.
5. V.S. Vyas, "Norms and and Efficiency in Agricultural Marketing", Seminar on Marketing and Agricultural Products, Bombay, **Indian Society of Agricultural Economics,** 1965. p.138-146.
6. Uma J. Lele, **Foodgrain Marketing in India,** *op.cit.*
7. R.W. Junior Cummings, *op.cit.* p.82.
8. Hanumantha Rao and K. Subbarao, *op.cit.*
9. A. Kumar and B.B. Singh, "Market Integration and Efficiency of Paddy Crop", **Indian Journal of Agricultural Economics**(October-December 1979); C.D. Deole, and others, "An Analysis of Wheat Price Differences in Regulated Markets", **Indian Journal of Agricultural Economics** October-December 1979, p.207.
10. Uma J. Lele, **"Traders of Sholapur,"** in Mellor Jr. et.al. (eds.) **Developing Rural India** : New York, Cornel University Press, 1968, pp.257-262.
11. S.S. Acharya and N.L. Agarwal, **Agricultural Marketing in India** Delhi: Oxford and IBH Publishing Company Private Limited, 1987. p. 307.
12. B.D. Joshi, **Working of Regulated Markets in India,** *op.cit.*
13. Z.Y. Jasdanwala, *op.cit.*, p.216.

VI

THE FINANCIAL ANALYSIS

This chapter discusses the income aspect of Agricultural Produce Market Committees (APMCs) of Kheda district. The chief source of income of the APMCs is (i) market fee and (ii) licence fee.

(i) *Market Fee*

The market committee has the statutory power to levy market fee on the notified crops that are transacted in the notified area. The market fee is levied in the whole of market area. In the State of Gujarat, the rate of market fee is 30 paise per hundred rupee worth of the agricultural produce. The market fee is collected from the trader only. The fee is not levied more than once on the same commodity, when resold. The market committee has the powers to prosecute the trader who evades the market fee.

The collection of market fee depends on the price level and on market arrival. In the market, a small rise in the prices of the notified crops brings about a big rise in the collection of market fee. The reverse is the case when there is a fall in prices. Similarly, for a given level of price, increase in market arrival results into increased income by way of market fee and vice versa. Table 6.1 shows the market fee collected during 1975-76 to 1984-85 in Kheda district of Gujarat.

During the period under study, the total market fee collected amount to Rs.206.32 lakhs. The contribution of market fee to the total receipt was 51.42 per cent in 1974-75. Between 1976-77 and 1979-80 the market fee collection fell down to 44 and 47 per cent. This may be due to the fall in the prices. In 1980-81, the revenue went up to 55.55 per cent of the total receipts. However, during the study period, the maximum amount of market fee collection was found in the year 1983-84. The amount of market fee collected was Rs.44.97 lakhs which was 70.11 per cent of the total receipts. The increase of the market fee collection in 1983-84 may be due to the rise in the market arrival in that year. Thus price level and market arrival are the two main determinants of the market fee collection.

TABLE 6.1: Market Fee Collected During the Period 1975-76 to 1984-85 in (APMCs) of Kheda District

(Rupees in lakhs)

Year (1)	Total Receipts (2)	Market Fee Collected (3)	Percentage of (3) to (2) (4)
1975-76	19.19	9.87	51.42
1976-77	19.88	8.90	44.78
1977-78	20.71	9.89	47.78
1978-79	23.48	11.15	47.50
1979-80	25.99	12.9	49.99
1980-81	27.78	15.43	55.55
1981-82	36.45	21.09	57.84
1982-83	46.73	28.88	61.80
1983-84	64.13	44.97	70.11
1984-85	63.76	43.15	67.68
Total	348.10	206.32	59.27

A multiple regression model is used to estimate the influence of these two factors. The weighted price indices of paddy, wheat and bajra (from 1975-76 to 1984-85) are taken as one independent variables and the market arrival (from 1975-76 to 1984-85) of the same crops namely, paddy, wheat and bajra is taken as the second independent variable. The other notified crops are excluded as their arrivals are very insignificant.

Then the equation is:

$$F_m = a + b_1 x_1 + b_2 x_2 + \mu$$

where

F_m = market fee
x_1 = price index
x_2 = market arrivals

b_1 and b_2 are the parameters to be estimated. The results are:

$\hat{\alpha}$	$\hat{\beta}_1$	$\hat{\beta}_2$	R^2
-3771724.8	30863.47	0.2164	0.94967*
	(8.354)	(1.167)	

* significant at 1% level.

With one per cent point increase in the weighted price index, the market fee collection increases by Rs.30863.47 and with one quintal increase in the market arrival, the market fee collection increases only by Rs.0.2164 at a given price index.

Barring the first two years of the study period as shown in Table 6.2, majority of APMCs fall in income group of Rs.100,000 - Rs.300,000. During 1981-82 to 1984-85, there was no APMC in the income group of Rs.20,000 - Rs.50,000. After 1982-83, some APMCs are found in the income group of Rs.500,000 and above. This shows the growth of APMCs of Kheda district in terms of the collection of market fees.

The overall performance of APMCs in terms of market fee collection is studied crop-wise with the help of tables 6.3. This table shows the share of selected crops in the total amount of market fee collection in APMCs of Borsad, Khambhat, Kapadvanj and Mehmedabad. In Borsad APMC, for all the years except one year of the study period, the market fee collection from paddy, rice, wheat and bajra was almost 50 per cent of the total collection. In 1982-83, however, nearly 70 per cent of the total collection was from these crops. In Khambhat APMC, the share of paddy, rice, wheat and bajra in the total market fee collection got reduced after the year 1980-81. In Kapadvanj APMC, the share of the selected crops viz. paddy, rice, wheat, and bajra was nearly 20 per cent during the study period. However, in Mehmedabad APMC, the share of the selected crops in the total fee collection was nearly 80 per cent in almost all the years of the study period.

Table 6.3 also shows the concentration of market fee collection i.e. the percentage share of four selected APMCs for the period 1975-76 to 1984-85. So far as paddy is concerned, its share in the total fee collection is highest in Mehmedabad APMC and lowest share is found in Kapadvanj APMC for most of the years during the study period. In case of rice, its share is found highest in Borsad APMC and lowest in

TABLE 6.2: Classification of APMCs of Kheda District on the Basis of Market Fee Collection During 1975-76 to 1984-85

Class Size	1975-76	1976-77	1977-78	1978-79	1979-80	1980-81	1981-82	1982-83	1983-84	1984-85
Less than Rs.10,000	-	-	-	-	-	-	-	-	-	-
Rs.10,000-20,000	-	-	-	-	-	-	-	-	-	-
Rs.20,000-50,000	2	3	4	2	2	2	-	-	-	-
Rs.50,000-100,000	5	5	2	3	3	3	3	1	2	-
Rs.100,000-300,000	4	4	5	6	6	5	7	7	4	5
Rs.300,000-500,000	-	-	-	-	-	1	1	2	1	3
Rs.500,000 and above	-	-	-	-	-	-	-	1	4	3
Total	11	11	11	11	11	11	11	11	11	11

TABLE 6.3: Cropwise Percentage Share of Paddy, Rice, Wheat and Bajra in Four Selected Market Yards in Market Fee Collection During 1975-76 to 1984-85

Name of Market	Name of the Crop	1975-76	1976-77	1977-78	1978-79	1979-80	1980-81	1981-82	1982-83	1983-84	1984-85
Borsad	(1) Paddy	3.90	4.03	5.92	7.33	8.62	10.49	13.76	11.20	10.96	17.55
	(2) Rice	10.23	11.88	11.31	11.88	15.49	13.72	13.26	20.78	18.86	14.37
	(3) Wheat	19.16	5.69	9.74	10.52	5.41	7.58	8.00	12.98	4.30	6.85
	(4) Bajra	9.98	13.83	16.49	17.34	17.19	19.89	13.98	24.07	9.22	9.78
	Total	43.27	35.43	43.46	47.07	46.71	51.68	49.00	69.03	43.34	48.55
Khambhat	(1) Paddy	32.38	42.00	53.91	22.07	42.00	52.20	14.82	21.61	17.45	16.55
	(2) Rice	1.74	3.83	2.96	0.42	3.85	2.75	0.20	0.06	0.21	0.19
	(3) Wheat	13.52	12.87	7.03	5.79	6.89	5.01	3.03	4.56	3.89	3.29
	(4) Bajra	2.75	3.92	3.91	3.40	7.95	5.41	3.79	1.32	3.71	2.73
	Total	50.39	62.62	67.81	31.68	60.69	65.37	21.84	27.55	25.26	22.76
Kapadvanj	(1) Paddy	0.88	1.79	6.25	3.16	2.70	1.92	3.43	1.45	4.59	4.52
	(2) Rice	2.51	2.34	3.27	3.14	4.48	4.48	3.41	4.90	7.29	5.02
	(3) Wheat	6.07	2.24	3.43	4.42	5.09	5.19	4.92	8.92	6.37	5.59
	(4) Bajra	1.67	3.44	3.79	4.35	6.23	8.43	7.72	10.56	5.40	4.81
	Total	11.13	9.81	16.74	15.07	18.50	20.02	19.48	25.83	23.65	19.94
Mehmedabad	(1) Paddy	54.62	58.51	68.77	50.76	52.32	40.93	53.62	53.35	60.79	52.25
	(2) Rice	3.43	3.63	4.43	5.44	4.94	4.07	4.96	7.76	6.45	10.50
	(3) Wheat	14.85	17.07	12.62	14.98	13.48	16.67	16.62	18.26	12.87	18.56
	(4) Bajra	5.30	6.38	8.36	7.05	9.40	9.20	1.23	13.92	4.98	6.75
	Total	78.20	85.59	94.18	78.23	80.14	70.87	76.43	93.29	85.09	88.14

Khambhat APMC for most of the years during the study period. In case of wheat, the highest share was claimed by Mehmedabad APMC for most of the years and lowest share by Kapadvanj APMC during 1975-76 - 1979-80. From 1979-80 to 1984-85, Khambhat APMC has the lowest share. In case of bajra, the highest share was claimed by Borsad APMC and the lowest by Khambhat APMC for most of the years.

The figures of market fees during 1975-76 to 1984-85 were fitted in the linear trend. The following are the results with the origin at October 1975-76:

$\hat{\alpha}$	$\hat{\beta}$	R^2	F
-200668.12	411610.21 (5.97)	0.8167	35.63

Thus, the trend of market fee collection shows an increase of Rs.411610.21 per annum over the period under study. However, the annual growth of market fee is 19.95 per cent.

The graphical presentation of the trend in the market fee collection along with the actual market fee collection has been given in figure 17 (Appendix II). Looking to the graph, it is found that the actual value fluctuates around the trend line. Before 1977-78, the actual values remained above the trend. In 1977-78 and 1982-83 the actual values were equal to the trend value. Between 1977-78 and 1982-83 the actual values have remained below the trend value. After 1982-83 the actual values have remained above the trend value.

Table 6.4 shows that F value is significant at 1 per cent, hence the two variables, viz. arrival and prices together significantly affect the market fee. However, from 't' values, it can be observed that only price variable is significant. This could be due to multicollinearity (.60940) between two explanatory variables. Nevertheless, the sign of market arrival is positive and hence proper. Moreover, the two variables together explain 94.97 per cent variation in market fee.

(ii) *Licence Fee*

The second source of income for APMCs is levy of licence fee. Licence fees is levied on the market functionaries, such as traders, weighmen, brokers, hamals. Only licence holders can do business in regulated markets. Evasion of licence fee is punishable. Licence fees are fixed and so the amount of licence fee varies with the number of licences that are issued.

TABLE 6.4: Analysis of Variance of Market Fee Collection in APMCs of Kheda District

	DF	SS	MS	F
Regression	2	0.1625396 E14	8.12698 E12	66.038*
Deviation	7	0.86146237E12	1.2306605E11	
Total	9	1.7115422 E13		

* Significant at 1% level.

Table 6.5 shows the amount of licence fee collected during 1975-76 to 1984-85 in APMCs of Kheda district. The amount of licence fee collection has increased over the years. During the period under study, the total licence fee collected amounted to Rs.50.06 lakhs. The contribution of licence fee to the total receipt was 14.38 per cent during 1975-76 to 1984-85. The contribution of licence fees to the total income of the APMCs of Kheda district has gone down from 22.17 per cent in 1975-76 to 9.09 per cent in 1984-85.

The figures of licence fee collection during 1975-76 to 1984-85 were fitted in linear trends. The following are the results obtained with the origin at October 1975-76.

$\hat{\alpha}$	$\hat{\beta}$	R^2	F
404303.68	17521.74 (12.30)	0.94981	151.41

Thus, the trend of licence fee collection shows an increase of Rs.17521.74 per annum over the period under reference. The annual growth rate of licence fee is 3.49 per cent.

Figure 18 (Appendix II) shows the trend in licence fee collection and the actual collection of licence fee during 1975-76 to 1984-85. It can be observed from the graph that before 1978-79, the actual licence fee collected was below the estimated trend value of the licence fee collection. Around 1978-79, and 1981-82, both of them tended to be equal. In between these two years, the actual licence fee collected was above the trend.

TABLE 6.5: Amount of Licence Fee Collected During 1975-76 to 1984-85 in APMCs of Kheda District

(Rupees in Lakhs)

Year (1)	Licence Fee Collected (2)	Total Receipts (3)	Percentage of (3) to (2) (4)
1975-76	4.25	19.19	22.17
1976-77	4.32	19.88	21.72
1977-78	4.42	20.71	21.36
1978-79	4.68	23.48	19.96
1979-80	5.11	25.99	19.66
1980-81	5.27	27.78	18.98
1981-82	5.37	36.45	20.28
1982-83	5.36	46.73	11.47
1983-84	5.49	64.13	8.55
1984-85	5.79	63.76	9.09
Total	50.06	348.10	14.38

Table 6.6 shows the financial aid by government institutions to the market committees as on 30th September 1985 in Kheda district. In 1984-85, the market committees of the district received Rs.10.13 lakhs worth financial assistance and Rs.21.45 lakhs by way of loan. The capital was worth Rs.14.51 lakhs and the interest that the committees paid was 6.23 lakhs. The unpaid loan was Rs.1.95 lakhs. Kapadvanj market committee received financial assistance worth Rs.5.68 lakhs which is the largest among the amounts received by all the market committees in 1984-85. Petlad committee received the lowest financial assistance (Rs.0.56 lakhs) in the same year. However, Balasinor, Khambhat, Anand and Nadiad APMCs did not receive any financial assistance. Khambhat market committee had taken Rs.4.99 lakhs by way of loan in 1984-85 which was the highest amount among the loans taken by the market committees. Khambhat APMC had the largest amount of permanent reserves of Rs.32.97 lakhs and Balasinor APMC had the smallest amount of reserves of Rs.3.06 lakhs among all market committees.

Conclusion

APMCs derive their major income from market fee and licence fee. Of the two, market fee is relatively more important. Market fee and licence fee collection in 1975-76 was Rs.9.87 lakhs and Rs.4.25 lakhs respectively. Both of them increased to Rs.43.15 lakhs and Rs.5.79 lakhs in 1984-85. The growth rates of market fee and licence fee were 19.95 per cent and 3.49 per cent respectively. In the market fee collection, of the two variables viz. prices and arrivals, the first one is found more important than the second one.

TABLE 6.6: Financial Aid by Government Institutions to Mark^t Committees of Kheda District as on 30th September 1985

(Rs. in lakhs)

Sr. No.	Name of APMCs	Financial assistance to market Committee	Loan received by market Committee	Loan Repayment		Unpaid Loan	Permanent Reserves as on 30th September
				Capital	Interest		
1.	Umreth	0.58	0.67	0.67	0.27	-	5.42
2.	Balasinor	-	1.52	0.88	0.63	0.64	3.06
3.	Khambhat	-	4.99	-	-	-	32.97
4.	Kapadvanj	5.68	1.00	1.00	0.17	-	26.96
5.	Mehmedabad	0.66	1.91	1.88	0.88	0.03	9.54
6.	Anand	-	1.35	1.35	0.37	-	26.55
7.	Thasra	0.65	2.60	2.60	0.58	-	8.35
8.	Nadiad	-	3.04	3.04	1.04	-	27.08
9.	Borsad	1.00	1.50	1.50	0.46	-	22.09
10.	Petlad	0.56	1.00	0.47	0.70	0.53	3.94
11.	Matar	1.00	1.87	1.12	1.13	0.75	10.31
	1984-85	10.13	21.45	14.51	6.23	1.95	

VII

CONCLUSIONS AND POLICY IMPLICATIONS

From the analysis in the preceding chapters, the following conclusions can be drawn regarding the working of regulated markets in Kheda district:

(1) Regulated. markets are competitive in terms of number of buyers and sellers, buyer concentration, freedom of entry and exit and mobility of the produce.

Thus, the hypothesis:

"Regulated markets have the features and characteristics of a competitive market" is proved valid in markets of Kheda district.

(2) Using the variance and correlation analysis of prices in different markets, it is observed that the selected markets viz. Borsad, Khambhat, Kapadvanj and Mehmedabad are spatially integrated.

Hence, the hypothesis:

"Market regulation creates integration" is found valid in markets of Kheda district.

(3) Market stability has been examined with the help of coefficient of variation of prices in different selected markets over a period of ten years. The findings are that Borsad is the most unstable market and Mehmedabad the most stable one. If ranked in descending order, they are Mehmedabad, Kapadvanj,

Khambhat, and Borsad.

Thus, the hypothesis:

"Market regulation provides stability of market in terms of prices" does not hold good in markets of Kheda district.

(4) Cyclical variation in prices are studied by fitting a linear trend of prices and examining the cyclical fluctuations around this trend. Cropwise cycles around the trend in different crops of different markets are discussed in Chapter V. In general, it can be said that the severity and length of the cycles differ.

(5) In case of cost of buying and selling, both traders and farmers experience cost economies, if they purchase and sell in regulated market yards.

Thus, the hypothesis:

"The users of the regulated market yards enjoy cost economies" is found valid.

(6) So far as the prices received by the farmers belonging to different size groups for different crops in different markets are concerned, no clear picture emerges from the study.

(7) Market efficiency measured through market arrival show that the market arrival of some market yards is above the average arrival of all the regulated market yards of the district and the market arrival of some other market yards is below the district average. The first group includes Khambhat, Nadiad, Borsad, Mehmedabad and Matar. The second group includes Anand, Kapadvanj, Thasra, Petlad, Balasinor and Umreth. The same problem has been examined from another angle with the help of primary field survey. The finding of the study are as under:

In case of market arrival of paddy, the performance of Mehmedabad market is better than that of Borsad.

So far as wheat is concerned, the performance in terms of market arrival as a percentage of marketable surplus, the markets can be arranged in the descending order like this (1)

Khambhat, (2) Mehmedabad, (3) Kapadvanj and (4) Borsad. With regard to Bajra, if the markets are arranged in descending order of their arrival performance, the ranking can be like this, (1) Khambhat, (2) Kapadvanj, (3) Borsad and (4) Mehmedabad.

(8) Distance as a barrier to market arrival has also been examined in the present study. In general, it shows that distance affects farmers of all size groups in all markets except Khambhat. In other words, the market arrival as a percentage of marketable surplus is more from a nearby village than what it is from a distant village. This is true for marginal, small and large farmers.

(9) Analysis of non-economic variables shows that there are no problems with regard to disposal quickness, delay in payment, physical amenities, helpfulness of the market staff, corruption and crime, pilferages and wastes. The only two problems attracting the attention pertain to grading and market information/consciousness.

(10) In the income of the market committees, market fee and licence fee are important components. Of the two, market fee is more important than licence fee. Taking market fee as a dependent variable and arrival and price as independent variables, the linear regression was run at the district level. It shows that of the two, viz. arrival and price, the second one plays an important part in determining the income of the market committee.

From the above findings of the study, the picture that emerges is like this: Though the regulated markets are competitive and integrated, they are not stable. It is obvious also that competitiveness and integrability do not necessarily ensure stability. Among the market yards studied here, some are more unstable and some are less unstable. Instability of markets is reflected through price variations in the short-run (i.e. seasonal) and in the long-run. Long-run instability in prices might be emerging from, besides other reasons, variations in agricultural production resulting from good years and bad years, i.e. from droughts and bumper crops. This type of problems are beyond the perview of marketing studies. They belong to the development of agriculture and to the overall development of the area. However, short-run fluctuations in prices result from uneven supply of the foodgrains in the market during the year i.e. excess supply at the time of harvest and reduced supply in the remaining part of the year. These types of problems emerge from the lack of storage facilities, lack of information and need for finance on the part of farmers. Thus price fluctuations resulting into cycles of different length and intensity should be reduced by devising appropriate measures. As found in the present study, among the farmers, it is the marginal and small farmers who suffer most because they bring their produce for the sale at the time of harvest for two reasons: (a) lack of storage facility and (b) need for finance. Thus, to reduce market instability resulting from excess supply at the time of harvest it is necessary to provide warehousing facility to preserve their produce and to provide finance to make them free from distressed sales. Subbarao[1] has rightly pointed out that the present marketing reforms have over-emphasized on changing the forms of marketing and market regulation has remained restricted to elimination of predatory middlemen, excessive marketing margins and unfair trade practices. He feels that market regulation should be further extended to (a) credit (b) transport and freight (c) processing and (d) storage because market regulation has yielded better results where these constraints are not serious. Where these constraints do not exist, markets have operated with reasonable efficiency even in the absence of market regulation. Alternatively speaking, where these constraints are overriding, market regulation is not able to solve the problems created by these constraints and they have a powerful neutralizing effect on the improvements brought about by the regulation of commodity markets.

Thus credit, transport and storage facilities should be provided to the farmers in general and marginal and small farmers in particular so that on one hand, market stability can be restored and on the other hand, their economic interests can be protected.

Further, it is observed from the present study that when the farmers sell outside the regulated market yard, they sell at a less remunerative price. In this situation, efforts should be made to attract them to the regulated market yard through economic and non-economic incentives. Among the economic incentives, actual price received by them is relatively more and they receive timely payment without any difficulty in the form of bribe, corruption, false weighment etc.

Grading is another important aspect of marketing. It has not been introduced so far in the selected and other market yards of Kheda district. It provides incentive to farmers to produce and bring better quality produce to the market yard. It discriminates in favour of good quality produce against the poor quality produce. From the point of view of buyers also, it rather makes it easy for those who want to buy a better quality produce at a higher price. Thus, all efforts should be made to introduce grading in the marketing of agricultural crops. For this purpose, improved training on the part of marketing staff has been rightly emphasized by Sain, K.[2]

Further from the primary survey, it is found that All India Radio broadcast is not heard by most of the farmers. Here in the present analysis, this problem has been examined from the point of view of size groups of farmers and not from the point of view of the distance of the village from the market yard. This is because radio waves can move to any corner of the world and therefore distance is not a problem as it is in case of telegrams, telephones and postal services. Further, price of a radio might be a deterrant factor in purchasing it and there his economic conditions based upon his size group might be relevant. What is important to understand is that radio is an audio-instrument which is less attractive in these days of television which is a video-instrument. So what can be suggested here is that within the programmes on agriculture, this information on prices should be telecast on television so as to make them reach to the farmers.

Distance from the market is another factor which deters the farmer from bringing his produce to the regulated market yard. This problem can be solved with the help of the development of cooperative marketing societies. The major weaknesses of these marketing societies is that they are saddled with bureaucrats/procedures and are of non-viable size. Further, they are established at the institute of the State level to reach the plan targets.

The intervention of governmental agencies in the process of marketing has brought about noticeable changes in the costs and margins of different agencies involved in marketing. On the whole, marketing of foodgrains has become more efficient due to state intervention. This is because of the fact that the participation of government agencies has curbed speculative propensities of the traders. But despite the increase in marketing efficiency, operating cost of the governmental agencies is relatively higher as compared to that of the private trader. Therefore, there is an urgent need to conduct indepth analysis of different elements of their cost structure with a view to streamlining the procedures and thus economise on costs. In Gujarat, all such improvements are visible. However, the scope for bringing down the operating cost of governmental agencies can be explored in Gujarat state. As this problem falls out of the purview of the present study, it has not been touched here.

The possibility of introducing discriminatory charges for farmer belonging to different size group in the regulated markets of Gujarat can also be explored.

There are improvements and short-falls in the field of agricultural marketing as seen above. Some researchers and scholars feel that the reforms have fell short of the expectations. Here one is reminded of K. Subbarao's[3] remark that the task of the overall assessment of the improvement in marketing system is not easy because the studies on marketing are either crop specific or region specific. In other words, the generalisation is difficult and the overall view remain inconclusive.

NOTES

1. K. Subbarao, "Presidential address", *op.cit.*
2. K. Sain, "Measurement of Margins for Some Crops", **Agricultural Marketing,** 23(4), 1981, pp.5-10.
3. K. Subbarao, "Presidential Address", *op.cit.*

APPENDICES

APPENDIX I

PARITY PRICES

Kulkarni has attempted to estimate the price spread of groundnut between two regulated markets in Maharashtra with the help of oil prices prevailing in Bombay during 1958-59[1]. He has developed the concept of parity price. A small exercise on the lines of Kulkarni has been attempted and presented in this Appendix. The reason for putting it in Appendix being that the main study pertains to four selected market yards of Kheda district. Parity of price of the selected market yards with prices in terminal markets (like Ahmedabad) which are outside the Kheda district does not fit within the framework of the study. He has used the following equation to know whether the prices at two place are in parity or not.

$$L_{po} = (\alpha.B_o + \beta.B_c) - (\alpha.T_o + \beta.T_c) - C_o$$

$$= \alpha(B_o - T_o) + \beta(B_c - T_c) - C_o$$

where L_{po} is the oil millers parity price of groundnut (in Rs. per maund).

B_o= is the price of one maund of oil in Bombay

B_c= is the price of one maund of cake in Bombay

T_o= is the cost of transporting one maund of oil to Bombay

T_c= is the cost of transporting one maund of cake to Bombay

C_o= is the cost of purchasing and crushing one maund of groundnut pods

$\alpha =$ is the oil content in one maund of groundnut pods, and

$\beta =$ is the content of cake in one maund of groundnut pods.

This equation can be used to study market integration and thereby market competitiveness. The parity price can be estimated to know whether the actual price is more or less than the parity price and thereby to find out whether the markets are integrated or not.

In the present study, Ahmedabad has been taken as a terminal market and an attempt has been made to find whether groundnut-kernel prices in Kapadvanj market are in parity with Ahmedabad groundnut oil prices or not. The analysis is restricted to only Kapadvanj market because groundnut arrivals are recorded only in Kapadvanj market. Other markets like Borsad, Khambhat and Mehmedabad have negligible arrivals of groundnut and therefore are not recorded also. For this, monthly annual average prices of 1987-88 of groundnut oil in Ahmedabad and groundnut kernel in Kapadvanj have been used in the present study.

Kulkarni's equation has been accepted with some modifications. The modified equation is as under:

$$L_{po} = (\alpha A_o + \beta A_c) - (\alpha T_o + \beta T_c) - (\alpha A_o t_o + \beta A_c t_c) - (C_p + C_m)$$

$$= \alpha A_o + \beta A_c - \alpha T_o - \beta T_c - \alpha A_o t_o - \beta A_c t_c - C_p - C_m$$

$$= \alpha A_o - \alpha A_o t_o - \alpha T_o + \beta A_c - \beta A_c t_c - \beta T_c - C_p - C_m$$

$$= \alpha(A_o - A_o t_o - T_o) + \beta(A_c - A_c t_c - T_c) - (C_p + C_m)$$

where

L_{po} is groundnut parity price of groundnut kernels in Rs. per quintal.

A_o is price of groundnut oil in Ahmedabad

A_c is price of groundnut cake in Ahmedabad

α is the oil content from one quintal of groundnut kernel.

β is the cake content from one quintal of groundnut kernel.

T_o is the cost of transporting of oil (.41 quintal) extracted from one quintal groundnut kernel from Kapadvanj to Ahmedabad.

T_c is the cost of transporting cake (.57 quintal) extracted from one quintal groundnut kernel from Kapadvanj to Ahmedabad.

t_o is the tax on groundnut oil worth Rs.1/-.

t_c is tax on oil cake worth Rs.1/-.

C_p is the cost of purchasing one quintal of groundnut kernel, including tax, tolai, and weighment charges.

C_m is the cost of crushing one quintal groundnut kernel.

Appendix Table 1: Values of the Constant Terms in the Equation of Parity Prices

Market	α	T_o (Rs.)	β	T_c (Rs.)	t_o (Rs.)	t_c (Rs.)	C_p (Rs.)	C_m (Rs.)
Kapadvanj	0.41	6.66	0.57	4.00	0.05	0.05	42	15

where, α is the oil content from one quintal of groundnut kernel.

β is the cake content from one quintal of groundnut kernel.

T_o is the cost of transporting oil (0.41 quintal) extracted from one quintal groundnut kernel from Kapadvanj to Ahmedabad.

T_c is the cost of transporting cake (0.57 quintal) extracted from one quintal groundnut kernel from Kapadvanj to Ahmedabad.

t_o is the tax on groundnut oil worth Rs.1.

t_c is the tax on oil cake worth Rs.1.

C_p is the cost of purchasing one quintal of groundnut kernel including tax, tolai and weighment charges.

C_m is the cost of crushing one quintal groundnut kernel.

Appendix Table 2: Prices of Oil and Cake in Ahmedabad in the Months of 1987-88

(*Rupees per quintal*)

Month	Groundnut oil (A_o)	Groundnut Cake (A_c)
October	2466.3	311.3
November	2636.4	343.8
December	2563.7	373.7
January	2377.0	371.1
February	2227.7	366.3
March	2220.0	353.7
April	2213.5	329.9
May	2116.1	302.5
June	2158.7	290.2
July	2311.9	300.8
August	2138.1	225.3
September	1936.2	186.2

Source: GROFED.

Appendix Table 3: Parity Price of Groundnut Oil and Local Price of Groundnut Kernels in Kapadvanj in the Months of 1987-88

Months	Parity Price	Local Price	Price spread
October	1067.18	1570.05	-502.87
November	1151.04	1875.00	-723.96
December	1138.91	1700.00	-561.09
January	1064.78	1550.00	-485.22
February	1004.03	1850.00	-845.97
March	994.21	1500.00	-505.79
April	978.79	1170.00	-191.21
May	926.02	1425.00	-498.98
June	935.94	1010.00	-74.06
July	1001.35	1130.00	-128.65
August	892.78	1110.00	-217.22
September	792.97	1155.00	-362.03

The value of constant terms viz., T_o, T_c, t_o, t_c C_p and C_m are given in Appendix table 1. The values of constant terms were obtained from the oil-mills of Kapadvanj. The oil prices and cake prices prevailing in Ahmedabad during October 1987 to September 1988 were obtained from Gujarat Cooperative Oilseeds Growers' Federation Limited (GROFED). The groundnut kernel prices for the same period were obtained from Agricultural Produce Market Committee, Kapadvanj. The oil prices and cake prices are presented in Appendix table 2. The parity prices calculated with the equation mentioned above are presented in Appendix table 3 along with the local prices and the price spread between the parity price and the local price.

As shown in Appendix table 3, groundnut oil prices in Ahmedabad in all the twelve months of the year 1987-88 are not in parity with the groundnut kernel prices prevailing in Kapadvanj. Thus, Ahmedabad oil market is not integrated with Kapadvanj's kernel market.

Kulkarni also finds in his study that oil price of Bombay is not in parity with kernel price of Dhulia and Barsi. The reasons given by him for the lack of parity are:

(1) rise in local demand for oil due to establishment of Vanaspati factories in local areas especially Dhulia.

(2) High speculative activity in Barsi has pushed up the local prices of pods and oil millers in Barsi have crushed a very small

quantity of the market arrivals. Thus there is no commodity flow between the terminal market of Bombay and supplying markets of Dhulia and Barsi. The exercise of parity price is meaningful only when there is a commodity flow between two markets selected for the study and the supplying market caters only to the needs of the receiving market.

When the supplying market sends its produce to different markets, there will be different prices showing parity with different markets on different days or on the same day. The average of all of them need not show parity with the price of supplying market. Further the price spread may be influenced by local supply and demand conditions in local markets also.

In the present study the lack of parity between the oil prices of Ahmedabad and kernel prices of Kapadvanj might be explained on the lines of Kulkarni - i.e. lack of commodity flow between Ahmedabad and Kapadvanj. The other probable reasons for lack of parity include shortage of milling capacity,excessive profit margins and supply and demand conditions.

APPENDIX II

CYCLICAL RELATIVES OF PADDY PRICES

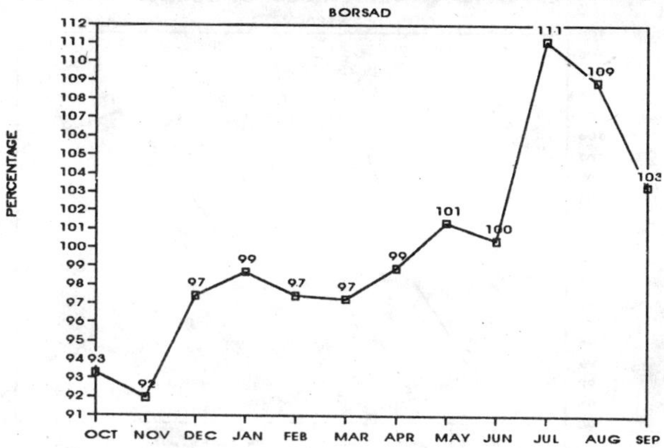

FIGURE-1

CYCLICAL RELATIVES OF PADDY PRICES

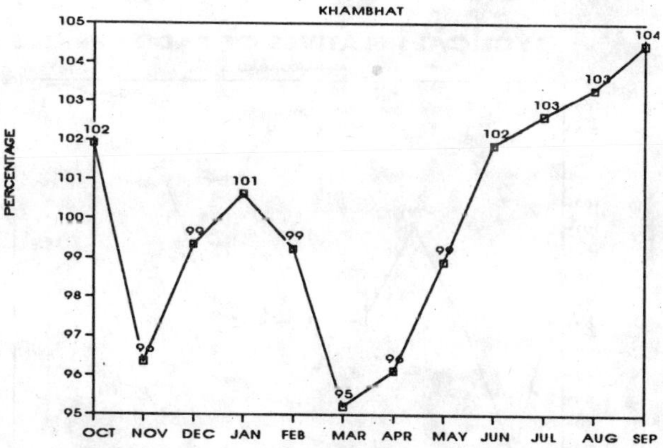

FIGURE-2

CYCLICAL RELATIVES OF PADDY PRICES

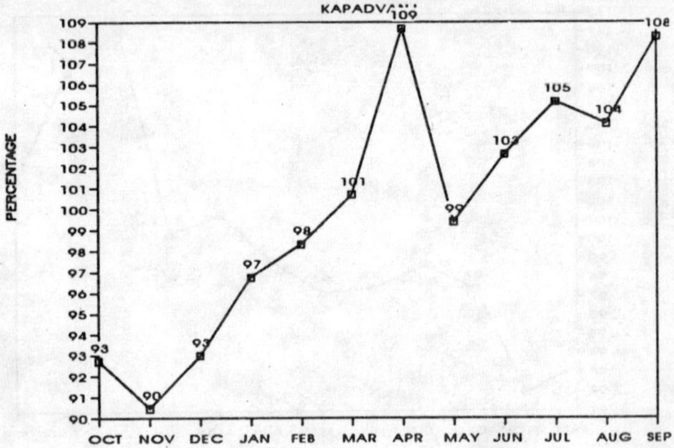

FIGURE-3

CYCLICAL RELATIVES OF PADDY PRICES

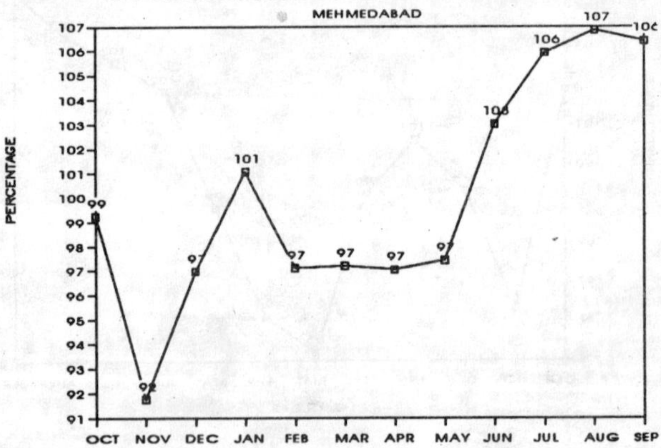

FIGURE-4

CYCLICAL RELATIVES OF RICE PRICES

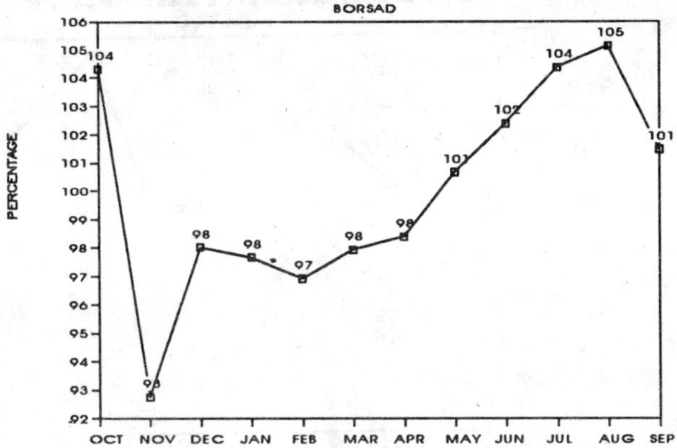

FIGURE-5

CYCLICAL RELATIVES OF RICE PRICES

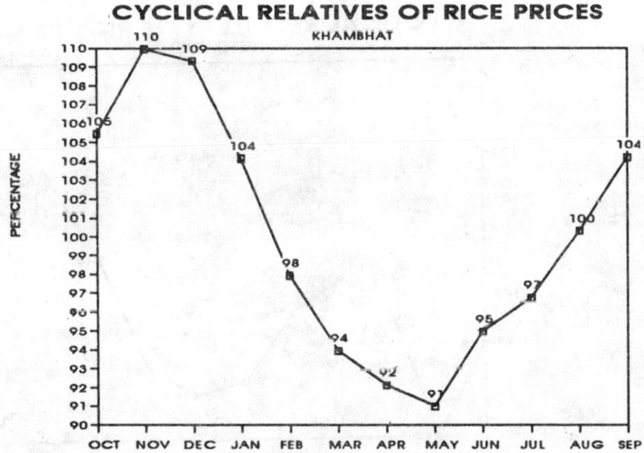

FIGURE-6

CYCLICAL RELATIVES OF RICE PRICES

KAPADVANJ

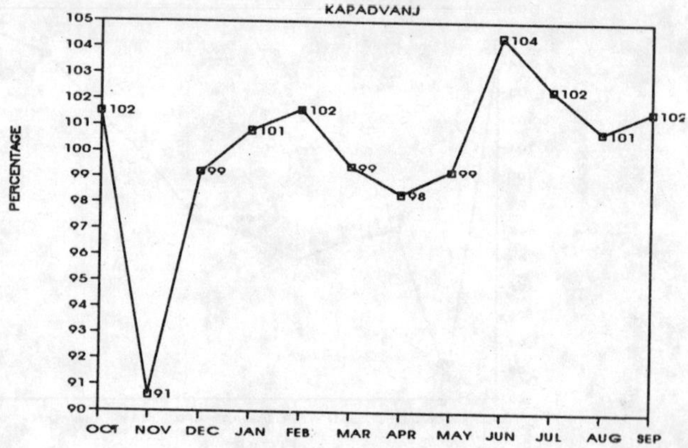

FIGURE-7

CYCLICAL RELATIVES OF RICE PRICES

MEHMEDABAD

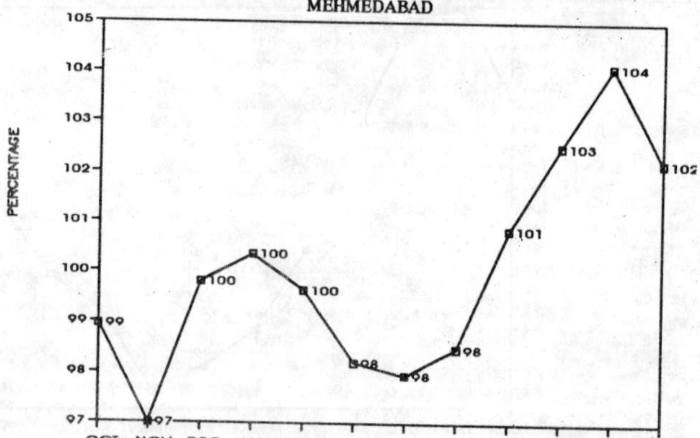

FIGURE-8

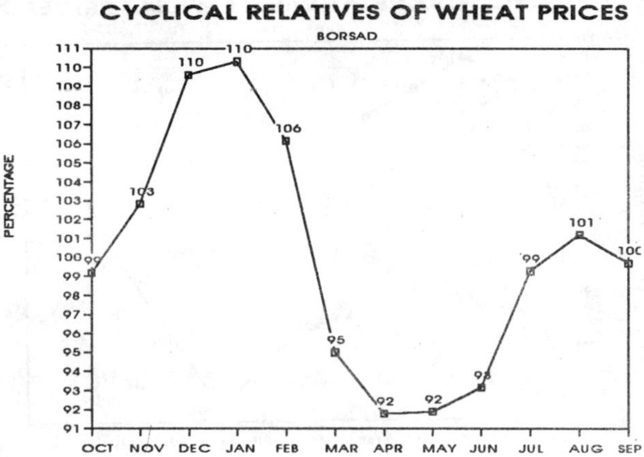

FIGURE-9

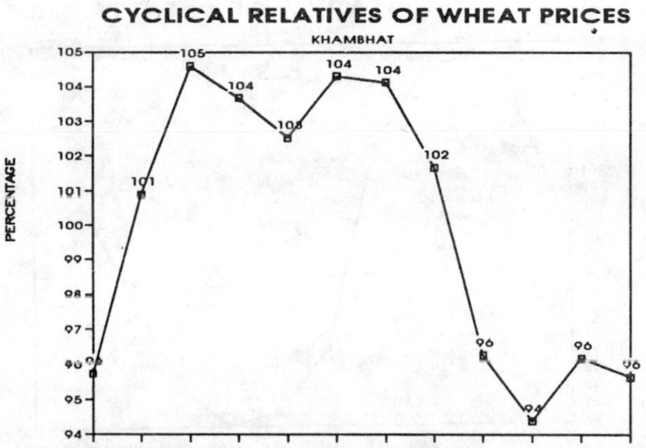

FIGURE-10

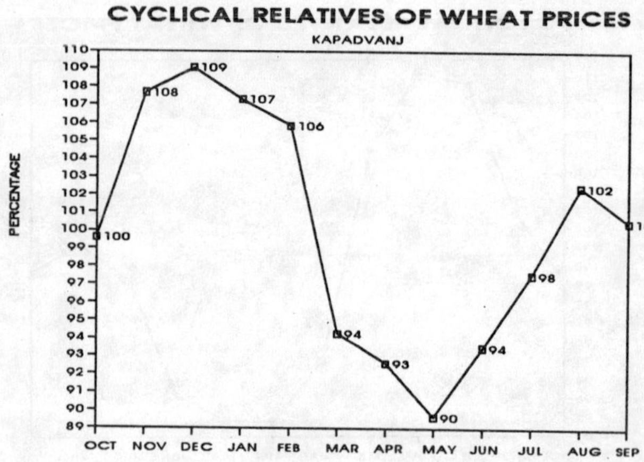

FIGURE-11

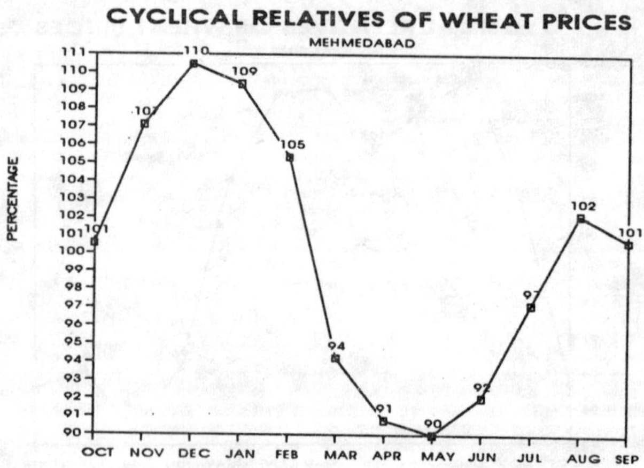

FIGURE-12

CYCLICAL RELATIVES OF BAJRA PRICES

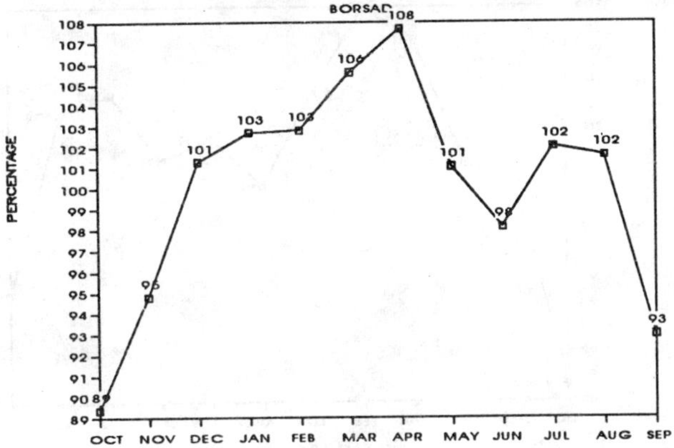

FIGURE-13

CYCLICAL RELATIVES OF BAJRA PRICES

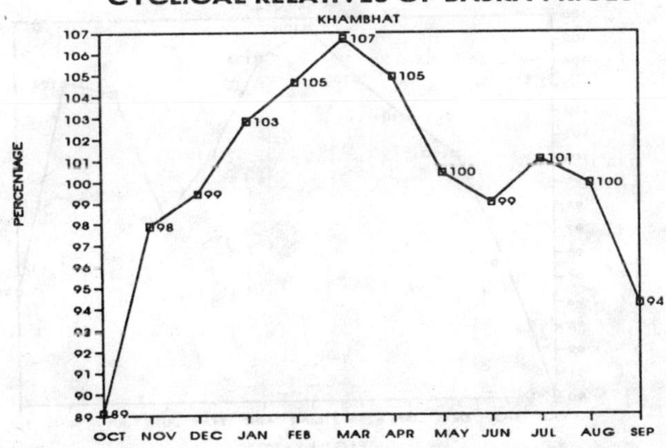

FIGURE-14

CYCLICAL RELATIVES OF BAJRA PRICES

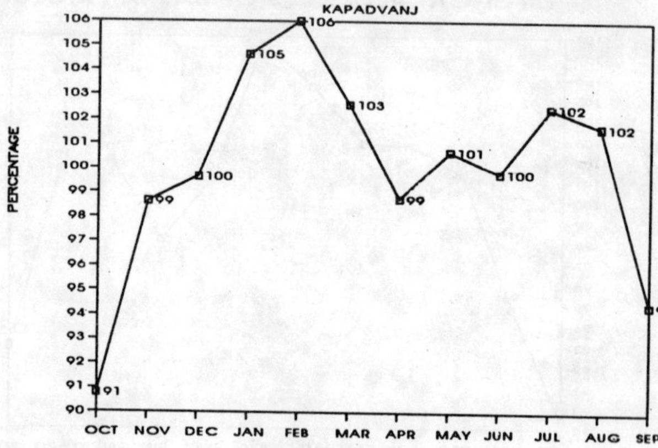

FIGURE-15

CYCLICAL RELATIVES OF BAJRA PRICES

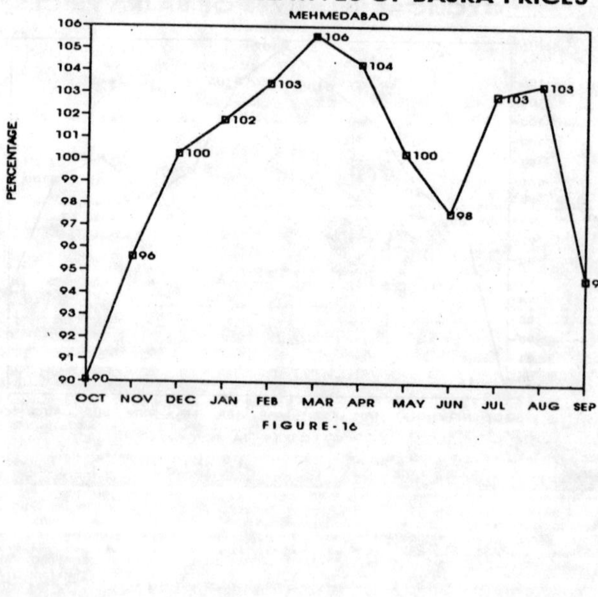

FIGURE-16

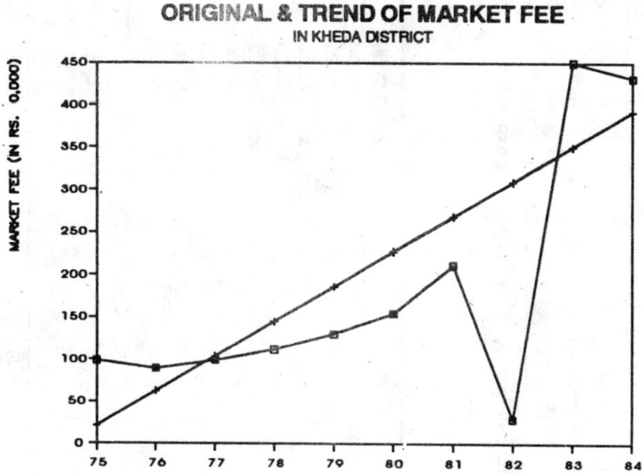

FIGURE-17

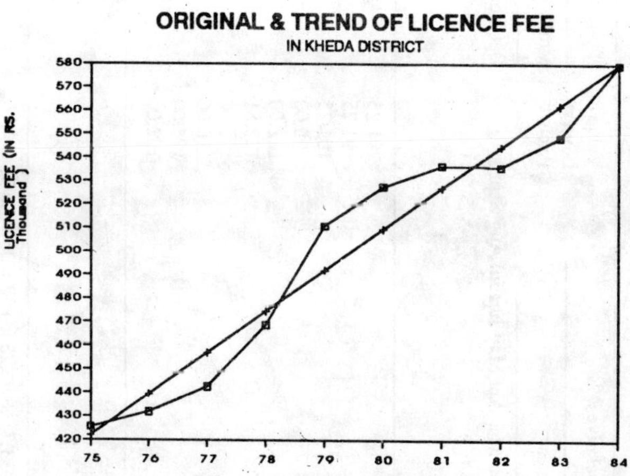

FIGURE-18

◻ ACTUAL + ESTIMATED

APPENDIX III

Seasonal Movement of Paddy Prices During 1975-76 to 1984-85

October to September Price	Lowest Price (L.P) relatives / Month of L.P. relatives												Value of Lowest Price	Highest Price (H.P) relatives / Month of H.P. relatives												Value highest relative	No.of mths. between the lowest and the highest	Seasonal increase
	Oct	Nov	Dec	Jan	Feb	Mar	Apl	May	Jun	Jul	Aug	Sep		Oct	Nov	Dec	Jan	Feb	Mar	Apl	May	Jun	Jul	Aug	Sep			
1975-76						x							91.25												x	125.00	5	37@
1976-77		x											91.25										x			149.50	7	64**
1977-78		x											101.25											x		121.25	8	20
1978-79		x											95.00												x	122.50	9	29
1979-80		x											122.90											x		140.00	8	14*
1980-81		x											122.81												x	180.25	9	47
1981-82	x												125.00												x	200.00	10**	60
1982-83	x												190.00											x		225.00	9	18
1983-84	x												175.00										x			280.00	8	60
1984-85							x						162.00										x			250.00	2*	54
																											75	403
Average (for 10 years)																											7.5	40.3
Range																											2-10	14-64

@ 1975-76: 125 — 91.25 = 33.75 ÷ 91.25 x 100 = 36.98 = 37

* Indicate minimum ** Indicate maximum

All the figures of the last column can be derived using the same method.

APPENDIX IV

RECOMMENDATIONS OF MASTER PLAN[2]

The farm production has increased over years in Gujarat State and the trend is likely to continue after Narmada Valley Project. Gujarat State Agriculture Marketing Board has got prepared a Master Plan which will develop regulated markets such that increased arrivals over a period of 15 years can be taken care of by them. The Master Plan has been submitted to Gujarat Government for its approval and to the Central Government for getting World Bank loan. It is a first major attempt to prepare a Master Plan at the state level to take care of future requirements.

The Plan recommendations include the following:

(1) To establish "Agricultural extension and agricultural marketing extension service centre". In this centre, farmers will be given extension education on improved marketing techniques and practices. This will help them to fetch higher price of their produce.
(2) To provide an expanded package of facilities like land, roads, water, power, fencing, auction platforms, primary cleaning and processing platforms, equipments for grading laboratories, scientific instruments for moisture testing and testing of oil content in oil seeds, sampling instruments, sieves etc.
(3) To construct godowns to store the produce on days when prices are not remunerative.
(4) To disseminate information of prices on television network.
(5) To integrate rural roads development with the development of market yards.
(6) To reduce wastes and marketing costs.
(7) To undertake research on agricultural marketing.
(8) To create special bank facilities right in the market yards so that farmers can obtain advances against the goods stored in the godown.
(9) To start a course on agricultural marketing in Agriculture

University.

Thus, the development plan proposes to develop 161 Agriculture Produce Market Committees and 352 market yards at a total estimated investment of Rs.163.29 crores. The objective behind the plan is (i) to increase farmers' income by bringing 90-95 per cent of the sale through market yard (2) to make the markets as "Instruments of Growth" and (3) to raise the status of some markets to National and International level. The plan is proposed to be implemented during the period of five years from 1988-89 to 1992-93.

NOTES

1. A.P. Kulkarni, "Price Spread for Groundnut in Two Regulated Markets", Seminar on Marketing of Agricultural Commodities, Bombay, **The Indian Society of Agricultural Economics**, 1965. pp.147-155.
2. **"Master Plan For The Development of Agricultural Produce Markets In Gujarat"**, Gujarat State Agricultural Marketing Board, Sector 22, Gandhinagar 382 022. The recommendations are based on the copy of the press note issued by the State Agricultural Marketing Board, Gandhinagar.

SELECTED BIBLIOGRAPHY

Abbot, J.C. "The Role of Marketing in the Development of Backward Agricultural Economies," *Journal of Farm Economics*, May 1962.

Acharya, S.S. *Agricultural Production Marketing and Price Policy in India : A Study of Pulses*, New Delhi: Mittal Publishers, 1988.

Acharya, S.S. and Agarwal, N.L. "Efficiency in Millet Marketing : A Case Study of Rajasthan", *Indian Journal of Agricultural Economics*, 39(3), Jaipur: Sukhadia Agricultural University, 1984.

Acharya, S.S. and Agarwal, N.L. *Agricultural Marketing in India*, Delhi: Oxford and IBH, 1987.

Agro-Economic Research Centre. *A Study of Production and Marketing of Bajra in Three Districts of Gujarat : Banaskantha, Sabarkantha, Kheda*, Vallabh Vidyanagar: Sardar Patel University, 1972.

Agro-Economic Research Centre, *Pace and Pattern of Marketing of Groundnut in Saurashtra Region of Gujarat*, Vallabh Vidyanagar: Sardar Patel University, 1972.

Bain Joe, S. *Industrial Organization*, New York: John Wiley, 1967.

Bapna, S.L. and Rao, K.R. *Supply and Price Outlook for Crops : A Study Based on Preharvest Market Information in Gujarat*, Delhi: Oxford and IBH, 1987.

Bhatt, B.D., Antaniasd, K.L., and Shiyani, R.L. "An Analysis of Arrivals and Prices of Important Vegetable Crops in Ahmedabad Regulated Market in Gujarat State," *Indian Journal of Agricultural Marketing*, 1 (1), June 1988.

Brahmbhatt, D.M. "Mango Marketing in Valsad District", Paper presented in the *Fifteenth Gujarat Economic Conference*, Surat, October - November 1983.

Bressler (Jr), G, ; Raymond and Richard, King A. *Market, Price and International Trade,* New York: John Wiley, 1970.

Brown, Dorris D. "The Use of Marketing Margin Data to Reflect Marketing Efficiency and to Develop Appropriate Market and Price

Policy : A Guide for Discussion", Bombay: *Indian Society of Agricultural Economics Seminar,* Series V, 1965.

Charan, A.S., Seetharaman, S.P., and Bapna, S.L. "Agricultural Marketing System in Gujarat: A Perspective", Paper presented at the *Fifteenth Gujarat Economic Conference,* Surat, October-November 1983.

Cummings, (Jr), R.W. *Pricing Efficiency in The Indian Wheat Market,* New Delhi: Impex India, 1967.

Dahl, Dale C. and Hammond, Jorome D.W. Market and Price Analysis, *The Agricultural Industries,* New York: McGraw-Hill, 1977.

Dantwala, M.L. "Principles and Problems of Agricultural Price Determination," *Indian Journal of Agricultural Economics,* June 1966.

——. "Problems Before the New Marketing Agencies," *Indian Journal of Agricultural Economics,* Volume 12, April-June 1957.

——. Report of the Expert Committee (This Committee was appointed by the Government of Maharashtra to enquire into Regulated Markets in Bombay in 1950) edited by K. Kulkarni, *Agricultural Marketing in India,* Volume 2, Bombay: Cooperators Book Depot, 1951.

Dave, P.K. "Risk and Uncertainty in Indian Agriculture" (Unpublished Ph.D. thesis), Gujarat University, 1980.

Deole, C.D. and others. "An Analysis of Wheat Price Differences in Regulated Markets," *Indian Journal of Agricultural Economics,* October-December 1979.

Dewey, Alice. *Peasant Marketing in Jawa,* New York: Free Press, 1969.

Dharm Narayan. *Distribution of the Marketed Surplus of Agricultural Produce by Size of Holding in India,* 1950-51, Asia Publishing House, New Delhi.

Directorate of Marketing, *Agricultural Marketing in India: Regulated Markets,* Volume I, New Delhi: Government of India, 1956.

Government of Gujarat. Gujarat Development Statistics; 1980, Gandhinagar: Centre for Monitoring Gujarat Economy, *Bureau of Economics and Statistics,* 1981.

——. Statistical Abstracts, Gandhinagar: Food and Coal Supplies Department, 1977-78.

Gupta, A.P. "Marketing of Agricultural Produce in India," *Commerce,* 130 (3343), 14 June 1975.

——. *Marketing of Agricultural Produce in India,* Bombay: Vora & Co., 1975.

Gujarat State Agricultural Marketing Board : Master Plan for the Development of the Agricultural Produce Markets in Gujarat, Gandhinagar: Government of Gujarat, 1987.

Gupta, V.K. and Mathur, D.P. *Changing Pattern of Wheat Production and Marketing,* CMA Monograph no. 67, Ahmedabad: Indian Institute of Management, 1977.

Harriss, Barbara, *Puddy and Rice Marketing in Northern Tamil Nadu,* Madras: Sangam Publishers, 1979.

——. *How to Study Agricultural Marketing and How Not to Study It.* Madras : Madras Institute of Development Studies, 41, (Working Paper no. 7.), 1981.

——. "Inaction, Interaction and Action: Regulated Agricultural Markets in Tamil Nadu, *Social Scientist,* 9(4), November 1980.

——. *State and Market,* New Delhi: Concept, 1983.

Indian Society of Agricultural Economics. Seminar on Emerging Problems of Marketing of Agricultural Commodities, Bombay: 1972.

Irwin, H.S. "Intangible Side of Agricultural Marketing : A Neglected Area of Research," *Journal of Farm Economics,* 14 (3), 1973.

Jasdanwalla, Z.Y. "Efficient Agricultural Marketing," *Economics and Political Weekly* (Supplement) 2 (53), 31 December 1977.

——. *Marketing Efficiency in Indian Agriculture,* Bombay: Allied Publishers, 1966.

Joshi, B.D. *Working of Regulated Markets in India,* Volume 2, Market Series no. 175, Nagpur : Department of Agriculture, Government of India.

Joshi, V.R. *Regulated Markets in Gujarat,* Vallabh Vidyanagar : Sardar Patel University, 1971.

Kaliyaperumal, P.N. "The Organization, Working and Performance of Regulated Markets in South Arcot District", (Unpublished Ph.D. Thesis), Madurai Kamaraj University, 1981.

Kahlon, A.S. and Balwinder Singh. *Marketing of Groundnut in Punjab.* Ludhiana : Punjab Agricultural University, 1968.

Kohls, R.L. and Uhl, J.N. *Marketing of Agricultural Products,* New York : Macmillan 5th edn. 1980.

Kulkarni, A.P. "Price Spread for Groundnut in Two Regulated Markets", 1965, Seminar Series V, Marketing of Agricultural

Commodities, Bombay : *Indian Society of Agricultural Economics*, 1965.

Kumar, A. and Singh, B.B. "Market Integration and Efficiency of Paddy Crop", In Thirty-ninth Annual Conference of the Indian Society of Agricultural Economics, Bangalore, December 18-29, 1979, Bombay : *Indian Journal of Agricultural Economics.*

Lalitha Agricons Private Limited. *Structural Analysis of Operations of Agricultural Produce Market Committee*, Unjha, Gandhinagar : Lalitha Agricons Private Limited, December 1986.

Lele, Uma. *Agricultural Policy in Developing Countries*, ed. Nurul Islam. London : Macmillan, 1974.

———. *Foodgrain Marketing in India : Private Performance and Public Policy,* Ithaca : Cornel University Press, 1971.

———. "Market Integration: A Study of Sorghum Prices in Western India," *Journal of Farm Economics*, February 1967.

———. "Traders in Sholapur" in Mellor Jr. et al. (eds), *Developing Rural India*, New York : Cornel University Press, 1968.

Madalia, V.K. "Functioning of Regulated Markets at Surat and its Impact," *Indian Journal of Agricultural Marketing*, 2, (1), June 1988.

Pandey, M.K. "Marketing Efficiency of Foodgrain in Gujarat," (Unpublished Ph.D. Thesis) , Gujarat University, 1982.

———, and Patel, M.S. *Marketing of Foodgrains in Gujarat* (Mimeo), Ahmedabad : Institute of Cooperative Management, 1979.

Patel, G.A. *Gujarat's Agriculture*, Ahmedabad: Overseas Book Traders, 1984.

Patel, R.J. and Desai, D.R. "Development of Agricultural Marketing in Gujarat," *Agricultural Marketing.* 15(4) January 1973, pp. 21-24.

Patel, S.K. "Problems of Tobacco Marketing in Kheda District", Paper presented at the *Fifteenth Gujarat Economic Conference*, Surat, October - November, 1983.

Purcell, Wayne, *Agricultural Marketing: System, Coordinator Cash and Future Prices*, Western Virginia: Reston Publishing Co., 1979.

Raju, V.T. and Kakadia, B.H. "Marketing of Groundnut in Rajkot District of Gujarat State," *Indian Journal of Agricultural Economics*, 3, July-September 1984.

——— and Von Oppen, M. *Marketing Margins and Price Correlations as Measures of Marketing Efficiency of Important Crops in Semi-Arid Tropical India* (Mimeo), International Crops Research Institute for the Semi-Arid Tropics, Hyderabad.

Ranade, C.G; Rao, K.H; and Shah, D.C. *Groundnut Marketing: A Study of Cooperative and Private Trade*, CMA Monograph No.92, Ahmedabad : Indian Institute of Management, 1982.

——; Singh, R.B; and Rao, Hanumantha. *Marketing Channels and Price Spread in Cotton*, CMA Monograph no. 83, Ahmedabad : Indian Institute of Management, 1979.

Rao, Hanumantha, and Subbarao, K. "Marketing of Rice in India," *Indian Journal of Agricultural Economics*, April-June, 1976.

——, and Subbarao, K. "Marketing of Rice in India: An Analysis of the Impact of Producers' Prices on Small Farmers," *Teaching and Research Forum no.18*, New York : Agricultural Development Council, May 1970.

Regulation of Agricultural Produce Market in Gujarat State, Gandhinagar : Gujarat Agricultural Marketing Board, 1988.

Reserve Bank of India. "Regulated Markets in India", June 1962, *Reserve Bank of India Bulletin.*

Rhodes, James V. *The Agricultural Marketing System*, Columbia : Grid Publishing, 1978.

Sain, K. "Measurement of Margins for Some Crops," *Agricultural Marketing*, 23(4), 1981.

Sau, R. "The Small Farmer and the Market with Special Reference to Rice and Wheat in India," *Economic Bulletin for Asia and Pacific*, 1975.

Subbarao, K. Presidential Address, *Indian Journal of Agricultural Marketing*, 2 (1), June 1988.

Thakur, D.S.; Chauhan, S.K.; and Sharma, K.D. "Efficiency and Weakness of Regulated Markets," *Indian Journal of Agricultural Marketing*, 2 (1), June 1988.

——. "Foodgrain Marketing Efficiency : A Case Study of Gujarat," *Indian Journal of Agricultural Economics*, 29(4), October - December 1977.

Thomson; Frederick; and Lundy. *Agricultural Marketing*, New York: McGraw-Hill, 1951.

——; ——; ——, and Foote, Richard Jay. *Agricultural Prices*, New York: McGraw-Hill. 1952.

Vyas, V.S. and Maharaja, M.H. "Factors Affecting Marketable Surplus and Marketed Supplies : A Study in Two Regions of Gujarat and Rajasthan," *Artha Vikas*, 2(1), January 1966.

——. "Norms and Efficiency in Agricultural Marketing," Bombay : *Indian Society of Agricultural Economies*, Seminar Series 5, 1965.

Wong, C.T. "Marketing of Locally Produced Vegetables in Hongkong," Some Studies of Fresh Fruit and Vegetable Marketing in Asia, ed. Herman Southworth, New York : *Agricultural Development Council*, 1974.

INDEX